Collins

AF813358

2027 GUIDE to the NIGHT SKY

Radmila Topalovic and Dominic Ford

Published by Collins
An imprint of HarperCollins Publishers
1 Robroyston Gate,
Glasgow G33 1JN
collins.reference@harpercollins.co.uk
collins.co.uk

HarperCollins Publishers
Macken House
39/40 Mayor Street Upper
Dublin 1
D01 C9W8
Ireland

In association with
Royal Museums Greenwich, the group name for the National Maritime Museum,
Royal Observatory Greenwich, Queen's House and *Cutty Sark*
www.rmg.co.uk

First published 2026

The publishers wish to acknowledge their deep debt of gratitude to two of their late astronomy authors: Storm Dunlop FRAS, FRMetS (1942-2025) for his stewardship of the Guides to the Night Sky and Night Sky Almanac and the enormous contribution of his friend and co-author the late Wil Tirion (1943-2024) for his expert star charts and diagrams. They both leave a valuable and ongoing legacy.

A catalogue record for this book is available from the British Library

ISBN 978 0 00 881575 2

10 9 8 7 6 5 4 3 2 1

Printed in Malaysia by Papercraft

If you would like to comment on any aspect of this book, please contact us at the above address or online.
e-mail: collins.reference@harpercollins.co.uk

MIX
Paper | Supporting responsible forestry
FSC® C007207
FSC www.fsc.org

Contents

Introduction

The aim of this Guide is to help people find their way around the night sky at any time of the year, by showing how the stars that are visible change from month to month, and by highlighting various events that occur during 2027. The objects and events described may be observed with the naked eye, or nothing more complicated than a pair of binoculars.

The conditions for observing naturally vary over the course of the year. During the summer, twilight may persist throughout the night and make it difficult to see the faintest stars. There are three recognized stages of twilight: civil twilight, when the Sun is less than 6° below the horizon; nautical twilight, when the Sun is between 6° and 12° below the horizon; and astronomical twilight, when the Sun is between 12° and 18° below the horizon. Full darkness occurs only when the Sun is more than 18° below the horizon. During nautical twilight, only the very brightest stars are visible. During astronomical twilight, the faintest stars visible to the naked eye may be seen directly overhead, but are lost at lower altitudes. As the diagram shows, full darkness persists for about six hours at midsummer in Sydney. Even at Christchurch, NZ (not shown), full darkness lasts about four hours. By contrast, Cape Horn is far enough south that at midsummer it never gets darker than nautical twilight.

Moonlight will affect the visibility of objects, providing a natural form of light pollution. At Full Moon, it may be very difficult to see some of the fainter stars and objects, and even when the Moon is at a smaller phase it may seriously interfere with visibility if it is near the stars or planets in which you are interested. A full lunar calendar is given for each month and may be used to plan night sky observations. It may also be useful to look at moonrise and moonset times for your particular location.

The celestial sphere

All the objects in the sky (including the Sun, Moon and stars) appear to lie at some indeterminate distance on a large sphere, centred on the Earth. This ***celestial sphere*** has various reference points and features that are related to those of the Earth. If the Earth's rotational axis (an imaginary line through the North and South Poles and the core of the Earth) is extended, for example, it points to the North and South Celestial Poles. Similarly, the ***celestial equator*** lies in the same plane as the Earth's

The duration of twilight throughout the year at Sydney and Cape Horn.

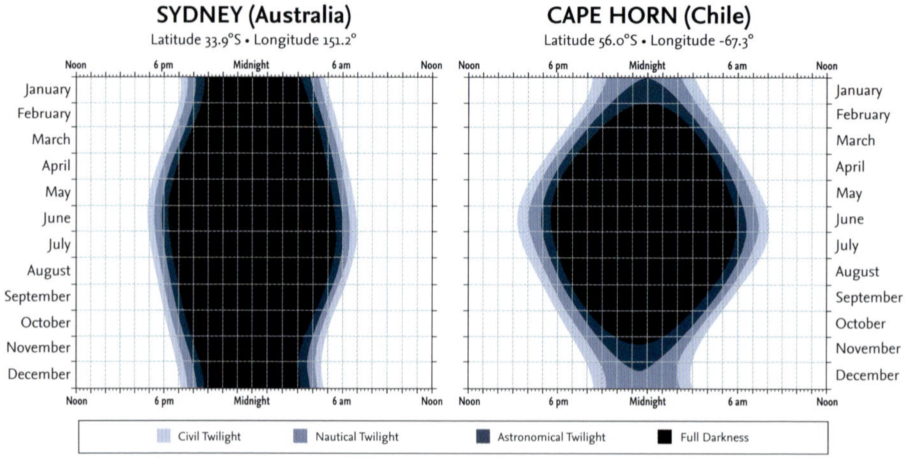

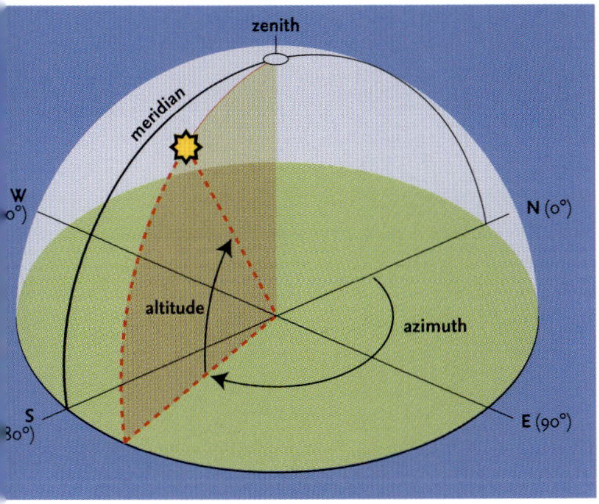

Measuring altitude and azimuth on the celestial sphere.

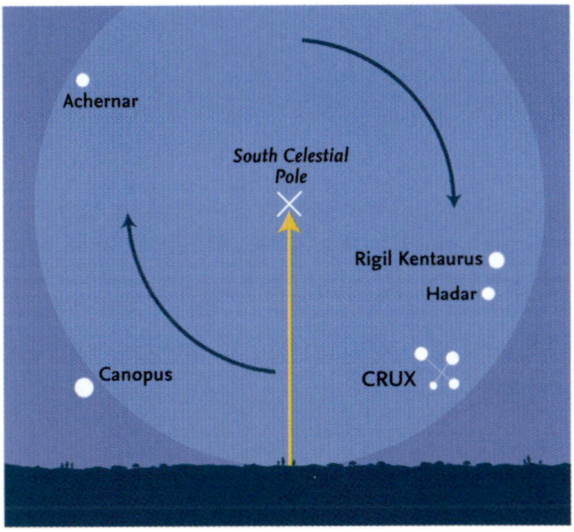

The altitude of the South Celestial Pole equals the observer's latitude.

equator; it divides the sky into northern and southern hemispheres. This Guide is written for use in the southern hemisphere, therefore the area of the sky that it describes includes the whole of the southern celestial hemisphere and those portions of the northern sky that become visible at different times of the year. Stars in the far north, however, remain below the horizon throughout the year, and are not included.

It is useful to know some of the astronomy terms for various parts of the sky. As seen by an observer, half of the celestial sphere is invisible at any point in time; these objects will be below the horizon. The point directly overhead is known as the ***zenith***, and the (invisible) one below one's feet as the ***nadir***. The line running from the north point on the horizon, up through the zenith and then down to the south point is the ***meridian***. This is an important invisible line in the sky, because objects are highest in the sky, and thus easiest to see, when they cross the meridian. Objects are said to ***transit*** when they cross this line in the sky.

In this book, reference is frequently made in the text and in the diagrams to the standard compass points around the horizon. The position of any object in the sky specific to the observer's location on Earth may be described by its ***altitude*** (measured in degrees above the horizon), and its ***azimuth*** (measured in degrees from north 0°, through east 90°, south 180° and west 270°). Experienced amateurs and professional astronomers also use another system of specifying locations on the celestial sphere called right ascension and declination, but here the simpler method will suffice.

The celestial sphere appears to rotate about an invisible axis, running between the North and South Celestial Poles. The location (i.e. the altitude) of the Celestial Poles depends entirely on the observer's latitude on Earth. At the South Pole (latitude -90° or 90°S), the South Celestial Pole (SCP) would be directly overhead (at the zenith, or at an altitude of 90°). The charts in this book are produced for the latitude of 35°S, so the SCP is 35° above the southern horizon. The fact that the SCP is fixed relative to the horizon means that all the stars within 35° of the pole are always above the horizon and may, therefore, always be seen at night, regardless of the time of year. This southern circumpolar region is an ideal place to begin learning the sky, and ways to identify the circumpolar stars and constellations will be described shortly.

The ecliptic and the zodiac

Another important line on the celestial sphere is the Sun's apparent path against the background stars – in reality the result of the Earth's orbit around the Sun. This is known as the ***ecliptic***. The point where the Sun, apparently

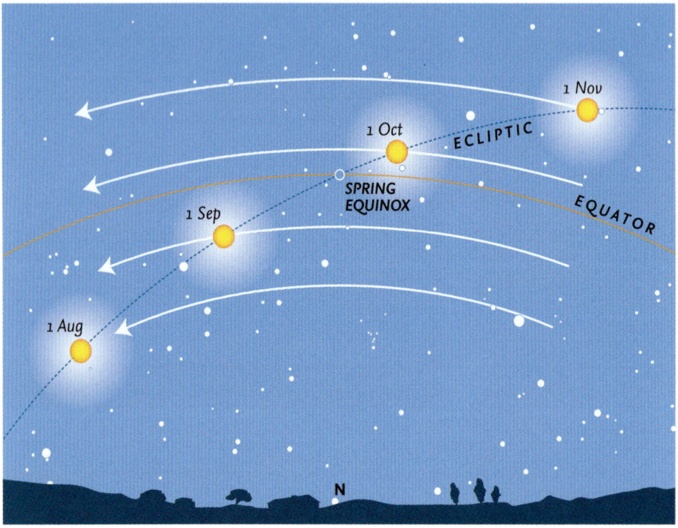

The Sun crossing the celestial equator in Spring.

moving along the ecliptic, crosses the celestial equator from south to north is known as the (southern) autumnal equinox, which occurs on 20 or 21 March. At this time and at the (southern) spring equinox, on 23 September 2027 (when the Sun crosses the celestial equator from north to south), day and night are almost exactly equal in length. The autumnal equinox is currently located in the constellation of Pisces; this means the Sun is moving through Pisces on this day. It is important in astronomy because it defines the zero point for the system of celestial coordinates (right ascension) not used in this Guide.

The Moon and planets are to be found in a band of sky all the way around the celestial sphere that extends 8° on either side of the ecliptic. This is because the orbits of the Moon and planets are inclined at various angles to the ecliptic (i.e. to the plane of the Earth's orbit). This band of sky is known as the zodiac and, when originally devised, consisted of twelve ***constellations***, all of which were considered to be exactly 30° wide. When the constellation boundaries were formally established by the International Astronomical Union in 1930, adjustments were made resulting in the ecliptic passing through thirteen constellations and the Moon and planets passing through several other constellations that are adjacent to the original twelve.

The constellations

Since ancient times, the celestial sphere has been divided into various constellations, most dating back to antiquity and usually associated with certain myths or legendary people and animals. The boundaries of the 88 constellations across the whole celestial sphere have been fixed by international agreement and their names (in Latin) are largely derived from Greek or Roman originals. Some of the names of the most prominent stars are of Greek or Roman origin, but many are derived from Arabic names. Many bright stars have no individual names and, for many years, stars were identified by terms such as 'the star in Hercules' right foot'. A more sensible scheme was introduced by the German astronomer Johannes Bayer in the early seventeenth century. Following his scheme – which is still used today – most of the brightest stars are identified by a Greek letter followed by the genitive form of the constellation's Latin name. An example is Sirius, known as α Canis Majoris (abbreviated α CMa). The Greek alphabet is shown on page 110, with a list of all the constellations that may be seen from latitude 35°S on page 109, together with abbreviations, their genitive forms and English names. Other naming schemes exist for fainter stars; these are not used in this book.

Asterisms

Apart from the constellations, certain groups of stars, which may form a part of a larger constellation or cross several constellations, are readily recognizable and have been given individual names. These groups are known as ***asterisms***, and the most famous (and well-known) seen from the southern hemisphere is 'the Southern Cross', the common name for the four brightest stars in the constellation of **Crux**. The names and details of some asterisms mentioned in this book are given in the list on page 110.

Magnitudes

The apparent brightness of a star, planet or other body is given in magnitudes (mag.). This is a mathematically defined scale where larger (positive) numbers indicate a fainter object. The scale extends beyond the zero point to negative numbers for very bright objects. (Sirius, the brightest star in the night sky is mag. -1.4.) Most observers are able to see objects as faint as about mag. 6, under very clear and dark skies.

The Moon

Although the daily spin of the Earth carries the sky from east to west (objects in the south rise in the east and set in the west), the Moon gradually moves eastwards relative to the background stars by approximately its diameter in an hour. This is equivalent to about half a degree across the sky. Its apparent eastward motion and change in phase are caused by its orbit around the Earth and changing position relative to the Sun.

Normally, in its orbit around the Earth, the Moon passes above or below the direct line between the Earth and the Sun at New Moon, or outside the area obscured by the Earth's shadow at Full Moon, as explained in the diagram on page 17. Occasionally, however, the three bodies are more or less perfectly aligned to give an ***eclipse***: a solar eclipse at New Moon or a lunar eclipse at Full Moon. During a solar eclipse, the Moon passes directly between the Earth and the Sun, obscuring the Sun for a brief period for part of the Earth. Depending on the exact circumstances, a solar eclipse may be merely partial (when the Moon does not cover the whole of the Sun's disk); annular (when the Moon is too far from Earth in its orbit to appear large enough to hide the whole of the Sun); or total. Total and annular eclipses are visible from very restricted areas of the Earth, but partial eclipses are normally visible over a wider area.

Somewhat similarly, at a lunar eclipse, the Moon passes through the shadow of the Earth. The Moon may pass through the lighter outer zone of the Earth's shadow, the ***penumbra***, called a penumbral eclipse, which is not generally perceptible to the naked eye; part of the Moon could pass within the darkest part of the Earth's shadow, the ***umbra***, in a partial eclipse; or it could move completely within the umbra in a total eclipse. Unlike solar eclipses, lunar eclipses are visible from large areas of the Earth.

Occasionally, as it moves across the sky, the Moon passes between the Earth and individual planets or distant stars, giving rise to an ***occultation***. As with solar eclipses, such occultations are only visible from restricted areas of the world.

The planets

The planets are named from the ancient Greek πλανήτης (planétēs) meaning wanderer, as they are always moving against the background stars. They are therefore treated in some detail in the monthly pages and information is given regarding when they are close to the Sun, other planets, the Moon or any of five bright stars that lie near the ecliptic. Such events are known as ***appulses*** or, more frequently, as ***conjunctions***. (There are technical differences in the way these terms are defined – and should be used – in astronomy, but these need not concern us here.) The positions of the planets are shown for every month on a special chart of the ecliptic.

The conditions of most favourable visibility depend on whether the planet is one of the two known as ***inferior planets*** – planets that orbit the Sun within the Earth's orbit (Mercury and Venus) – or one of the five ***superior planets***

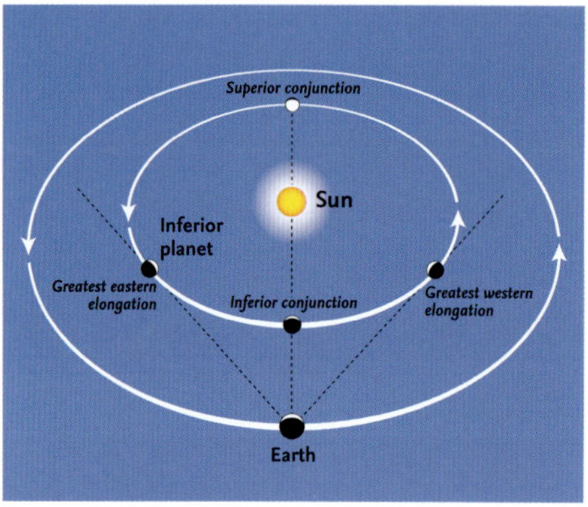

Inferior planet.

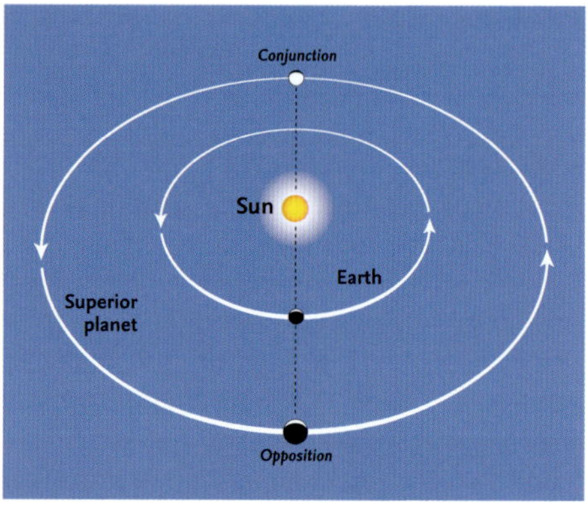

Superior planet.

– planets that orbit the Sun beyond the Earth. Of the latter, three (Mars, Jupiter and Saturn) are covered in detail; these are visible to the naked eye. Occasionally, details of the fainter superior planets, Uranus and Neptune, are included, and special charts are given for them on page 25.

The inferior planets are most readily seen at eastern or western elongation, when their angular distance from the Sun (separation from the Sun in degrees) is greatest. Superior planets are best seen at ***opposition***, when they are directly opposite the Sun in the sky and cross the meridian at local midnight.

Angular distance can be estimated by extending one hand at arm's length. The diagram opposite illustrates various angles that will be roughly equivalent to your handspan and the width of a clenched fist, and the separations between several stars in and around Orion.

Meteors

At some time or other, nearly everyone has seen a meteor – a 'shooting star' – as it flashed across the sky. The particles that cause meteors – known technically as 'meteoroids' – normally range in size from that of a grain of sand (or even smaller) to the size of a pea. ***Fireballs*** or ***bolides*** are very bright meteors (brighter than mag. -4) that are caused by objects up to 1 metre in size. Fireballs sometimes cause sonic booms that may be heard some time after the meteor is seen. On any night of the year there are occasional meteors, known as sporadics, that may travel in any direction. These occur at a rate that is normally between three and eight per hour.

Far more important, however, are meteor showers, which occur at fixed periods of the year, when the Earth encounters a trail of particles left behind by a comet or, very occasionally, by a minor planet (asteroid). Meteors of this type always appear to diverge from a single point on the sky, known as the radiant, and the radiants of major showers are shown on the charts. Meteors that come from a circular area 8° in diameter around the radiant are classed as belonging to the particular shower. All others that do not come from that area are sporadics (or, occasionally from another shower that is active at the same time). A list of the major meteor showers is given on page 31. Examples of meteors are shown on pages 32, 77 and 101.

Looking directly at the radiant is not the most effective way of seeing meteors. They are most likely to be noticed if one is looking about 40–45° away from the radiant position. This is approximately two hand-spans as shown in the diagram for measuring angles.

Other objects

In the late eighteenth century Charles Messier, a French astronomer, compiled a catalogue of objects while he was searching for comets. These nebulae, clusters of stars and galaxies, were given 'Messier numbers' (some already had names given several thousand years ago such as Praesepe – the Beehive Cluster). Some, such

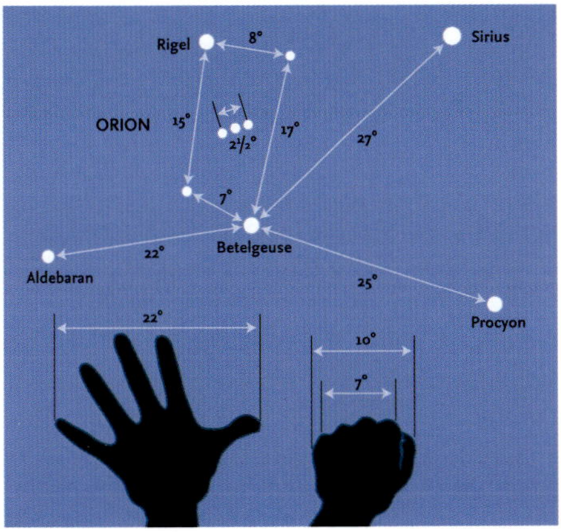

Measuring angles in the sky.

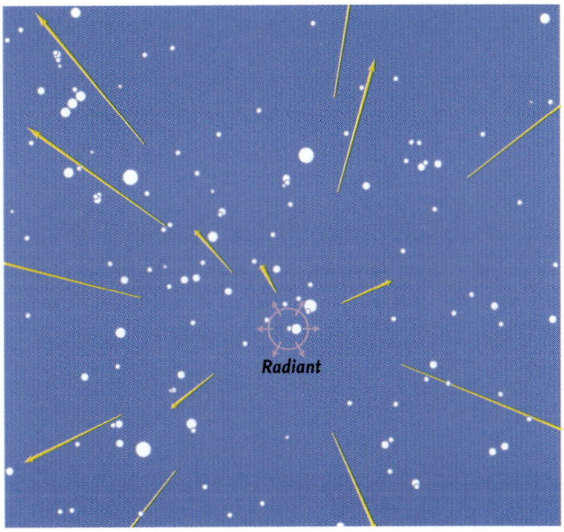

Meteor shower (showing the Geminids radiant).

as the Andromeda Galaxy, M31, and the Orion Nebula, M42, may be seen by the naked eye, but all those given in the list will benefit from the use of binoculars. Apart from galaxies, such as M31, which contain thousands of millions of stars, there are also two types of cluster: open clusters, such as M45, the Pleiades, which may consist of a few dozen to some hundreds of stars; and globular clusters, such as Omega Centauri, which are spherical concentrations of many thousands of stars. One or two gaseous nebulae, consisting of gas illuminated by stars within them, are also visible. The Orion Nebula, M42, is one, and is illuminated by a group of four stars, known as the Trapezium, which may be seen by using a good pair of binoculars.

Some interesting objects.

Messier / NGC	Name	Type	Constellation	Maps (months)
—	*47 Tucanae*	globular cluster	Tucana	All year
—	*Hyades*	open cluster	Taurus	Oct–Feb
—	*Melotte 111 (Coma Cluster)*	open cluster	Coma Berenices	Mar–Jun
M3	—	globular cluster	Canes Venatici	Apr–Jun
M4	—	globular cluster	Scorpius	Apr–Sep
M8	*Lagoon Nebula*	gaseous nebula	Sagittarius	Apr–Oct
M11	*Wild Duck Cluster*	open cluster	Scutum	May–Oct
M13	*Hercules Cluster*	globular cluster	Hercules	Jun–Aug
M15	—	globular cluster	Pegasus	Jul–Nov
M20	*Trifid Nebula*	gaseous nebula	Sagittarius	May–Oct
M22	—	globular cluster	Sagittarius	Apr–Oct
M27	*Dumbbell Nebula*	planetary nebula	Vulpecula	Jul–Oct
M31	*Andromeda Galaxy*	galaxy	Andromeda	Oct–Dec
M35	—	open cluster	Gemini	Dec–Mar
M42	*Orion Nebula*	gaseous nebula	Orion	Oct–Mar
M44	*Praesepe*	open cluster	Cancer	Jan–Apr
M45	*Pleiades*	open cluster	Taurus	Oct–Feb
M57	*Ring Nebula*	planetary nebula	Lyra	Jul–Sep
M67	—	open cluster	Cancer	Jan–May
IC 2602	*Southern Pleiades*	open cluster	Carina	Dec–Jul
NGC 2070	*Tarantula Nebula*	emission nebula	Dorado (LMC)	Aug–Jun
NGC 3242	*Ghost of Jupiter*	planetary nebula	Hydra	Jan–Jun
NGC 3372	*Eta Carinae Nebula*	gaseous nebula	Carina	Dec–Jul
NGC 4755	*Jewel Box*	open cluster	Crux	Jan–Sep
NGC 5139	*Omega Centauri*	globular cluster	Centaurus	Jan–Aug

The Southern Circumpolar Constellations

The southern circumpolar stars are the key to starting to identify the constellations. For anyone in the southern hemisphere they are visible at any time of the year, and nearly everyone is familiar with the striking pattern of four stars that make up the constellation of **Crux** (the Southern Cross) and also the two nearby stars ***Rigil Kentaurus*** and ***Hadar*** (α Centauri A and β Centauri, respectively). Both stars are actually triple star systems, one of the three stars of Rigil Kentaurus is called ***Proxima Centaurus***; it is the closest star to Earth, lying 4.2 light years away. This star has an Earth-sized planet, which orbits the star every 11.2 days. These circumpolar stars are visible throughout the year for most observers, although for observers farther north, the stars may become difficult to see as they are low on the horizon in the months of October and November.

Crux

The distinctive shape of the constellation of **Crux** is usually easy to identify, although some people (especially northerners unused to the southern sky) may wrongly identify the slightly larger False Cross, formed by the stars ***Aspidiske*** and ***Avior*** (ι and ε Carinae respectively) plus ***Markeb*** and ***Alsephina*** (κ and δ Velorum). The ***Coalsack Nebula*** (a vast dark cloud of obscuring dust up to 35 light years across) is readily visible on

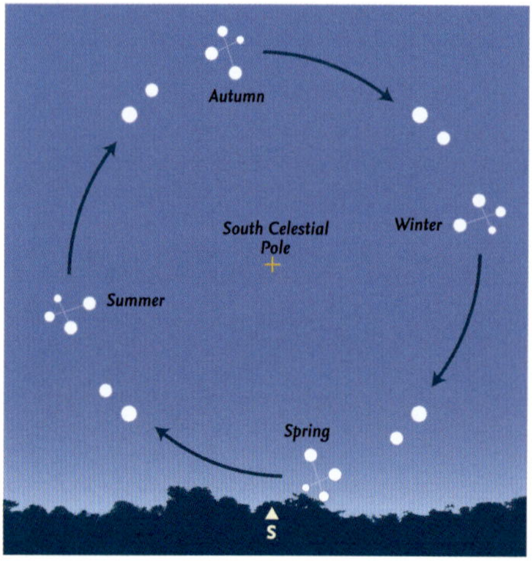

the southern side of Crux between ***Acrux*** and ***Mimosa*** (α and β Crucis).

A line through ***Gacrux*** (γ Crucis) and ***Acrux*** (α Crucis) points approximately in the direction of the South Celestial Pole (as described on page 5), which lies in the faint constellation of ***Octans***, crossing the constellations of ***Musca*** and ***Chamaeleon***. The basic triangular shape of Octans itself is best found by extending a line from ***Peacock*** (α Pavonis) in the constellation ***Pavo***, through β Pavonis, by about the same distance as that between the stars.

Centaurus

Although Crux is a distinctive shape, ***Centaurus*** is a large, rather straggling constellation, with one notable object: the giant, bright globular cluster ***Omega Centauri*** (the brightest and largest globular in the sky, consisting of around 10 million stars), which lies towards the north, on the line from ***Hadar*** (β Centauri) through ε Centauri. A more accurate method of locating the South Celestial Pole is to use the line from Crux and imagine a line at right angles and midway to the line between Hadar and ***Rigil Kentaurus*** – the intersection of these two lines identifies the SCP (see the diagram opposite).

Carina

Apart from the two stars that form part of the False Cross, the constellation of ***Carina*** is, like Centaurus, a large, sprawling constellation. It contains one striking open cluster, the ***Southern Pleiades***, and a remarkable emission nebula, the ***Eta Carinae Nebula***. The second brightest star in the sky (after Sirius) is ***Canopus***, α Carinae, which lies far away to the west.

The Magellanic Clouds

On the opposite side of the South Celestial Pole to Crux and Centaurus lie the two Magellanic Clouds, small irregular satellite galaxies of our Milky Way. The ***Small Magellanic Cloud*** (SMC) lies a distance of 200,000 light years from Earth; it sits to one side of the triangular

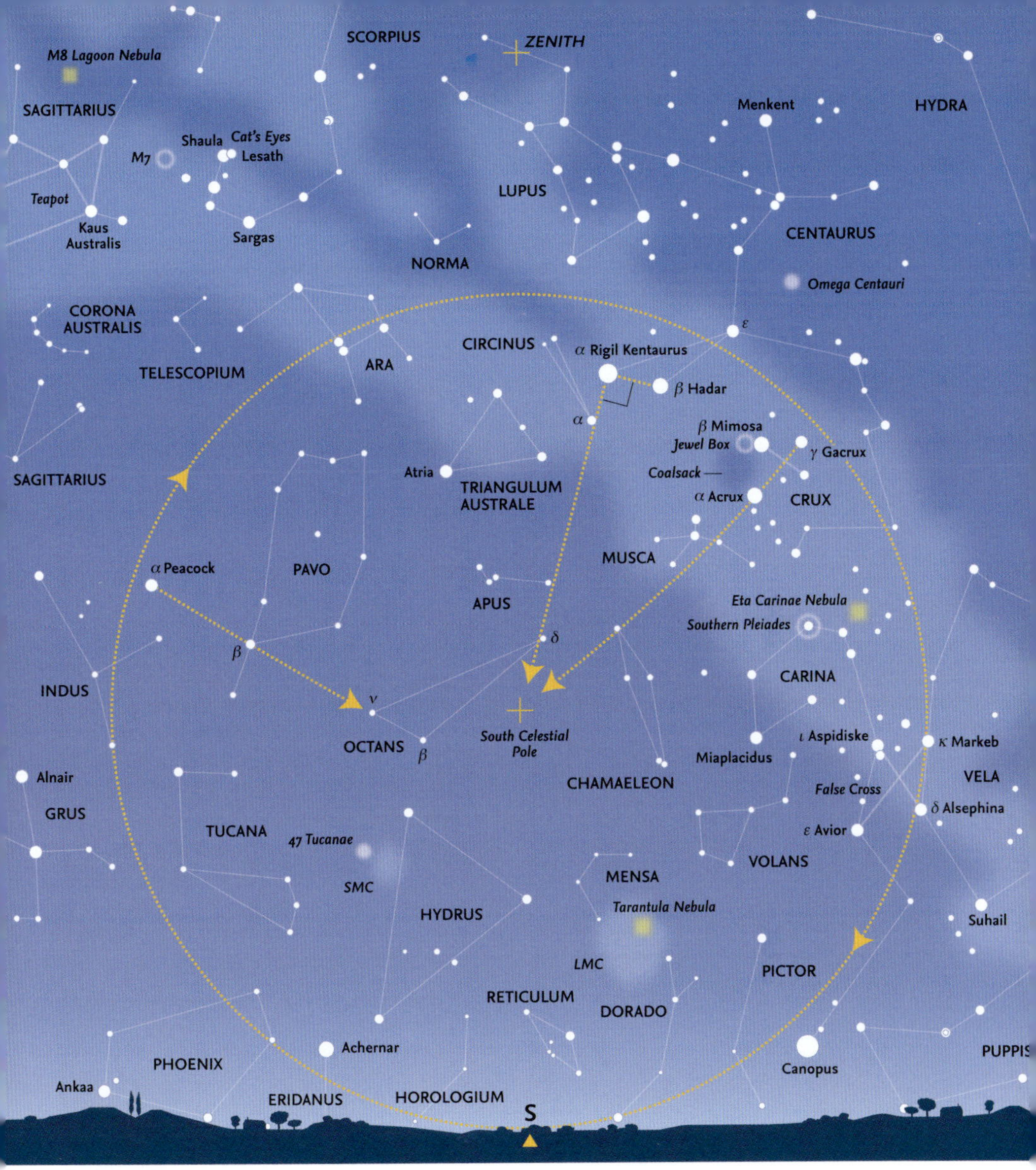

The stars and constellations inside the circle are always above the horizon, seen from latitude 35°S.

constellation of **Hydrus**, but it is actually within the constellation of ***Tucana***.

Nearby is another bright globular cluster, ***47 Tucanae***. Hydrus itself is also easily identified from the star ***Achernar***, α Eridani, the rather isolated brilliant star at the southern end of ***Eridanus***, which wanders a long way south, having begun at the foot of Orion.

The ***Large Magellanic Cloud*** (LMC) is around 40,000 light years closer to Earth than the SMC and it lies within the faint constellation of ***Dorado***. The LMC contains the large, readily visible ***Tarantula Nebula***, an emission nebula that is a major star-forming region. The LMC has an apparent size of around 10° in the sky, equivalent to 20 Moons across.

The Summer Constellations

The summer sky is dominated by several bright stars and distinctive constellations. The most conspicuous constellation is ***Orion*** (the Hunter), the main body of which has an hourglass shape. It straddles the celestial equator and is thus visible from anywhere in the world. The three stars that form the 'Belt' of Orion point to ***Sirius*** (α Canis Majoris), the brightest star in the night sky. ***Mintaka*** (δ Orionis), the westernmost star of the Belt, farthest from Sirius, points through ***Rigel*** in the direction of ***Achernar*** (α Eridani). A line from ***Alnitak*** (ξ Orionis, at the other end of the Belt) through ***Saiph*** (κ Orionis) points towards ***Canopus*** (α Carinae), the second brightest star in the night sky.

The bright red supergiant star ***Betelgeuse*** (α Orionis) lies below the belt, as does ***Bellatrix*** (γ Orionis). A line from Betelgeuse through Sirius indicates the general direction of the 'False Cross' in ***Vela*** and ***Carina***. A line from Bellatrix, through ***Aldebaran*** (α Tauri), past the 'V' of the ***Hyades*** cluster, points to the distinctive cluster of bright blue stars known as the ***Pleiades***, or 'the Seven Sisters'. Aldebaran is one of the five bright stars that may sometimes be occulted (hidden) by the Moon.

From the head of Orion, a line from ***Meissa*** (λ Orionis) through ***Elnath*** (β Tauri) leads to ***Capella*** (α Aurigae), and one from Betelgeuse through ***Alhena*** (γ Geminorum) points to ***Pollux*** (β Geminorum) of ***Gemini***. The slightly fainter second head of the twins, ***Castor*** (α Geminorum), lies to the northwest.

On the far eastern side of Orion sits the constellation of ***Leo*** (the lion), easy to find with its 'upside-down backwards question mark' (the 'head' of the lion).

Six bright stars in six different constellations: Capella (α Aurigae), Aldebaran (α Tauri), Rigel

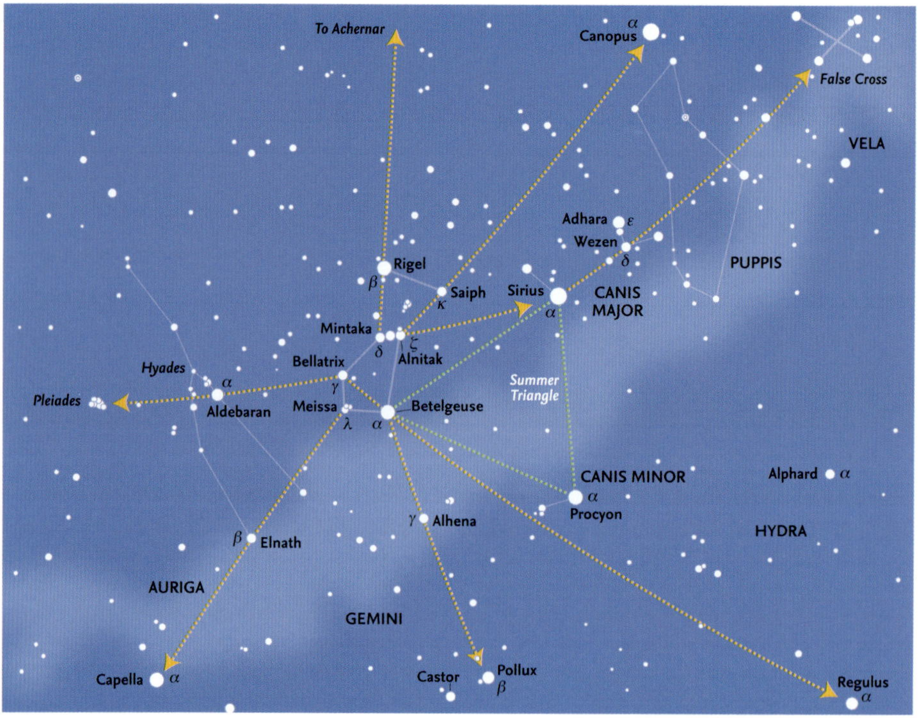

(β Orionis), Sirius (α Canis Majoris), ***Procyon*** (α Canis Minoris) and Pollux (β Geminorum) form what, for northern observers, is sometimes known as 'the Winter Hexagon'. An almost perfect equilateral triangle, 'the (southern) Summer Triangle', is formed by Betelgeuse (α Orionis), Sirius (α Canis Majoris) and Procyon (α Canis Minoris).

Several of the stars in this region of the sky show distinctive tints: Betelgeuse (α Orionis) is reddish, Aldebaran (α Tauri) is orange and Rigel (β Orionis) is blue-white.

The Autumn Constellations

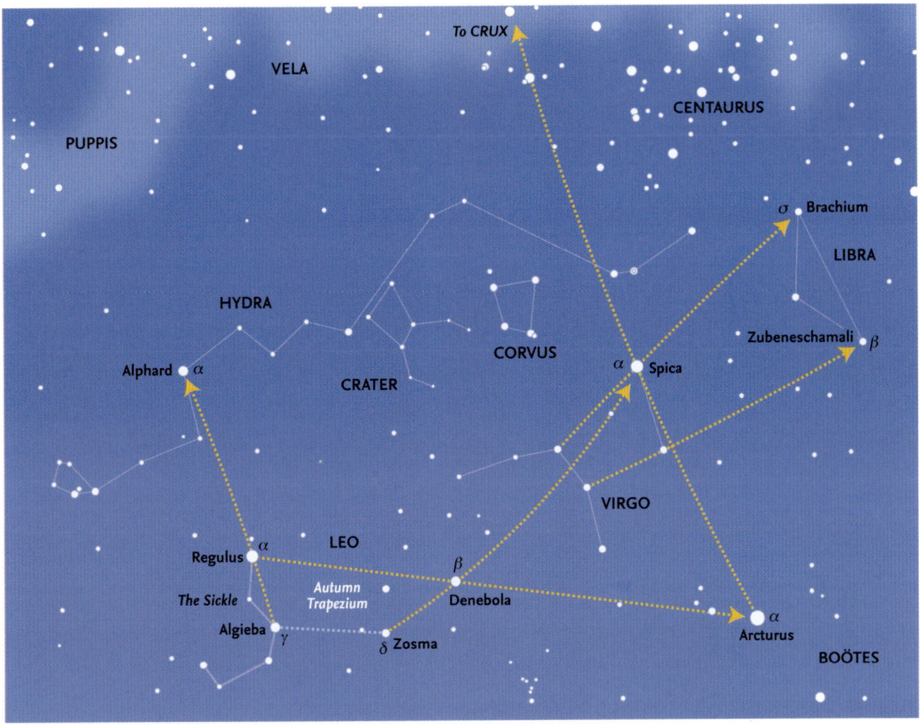

The most prominent constellation in the autumn sky is the zodiacal constellation of ***Leo***, and its brightest star, ***Regulus*** (α Leonis). Regulus is one of the five bright stars that could be occulted by the Moon. Regulus forms the 'dot' of the 'upside-down backwards question mark' of stars, known as 'the Sickle'.

A line extended from ***Algieba*** (γ Leonis), the second star in 'the Sickle', through Regulus points to ***Alphard*** (α Hydrae) the brightest star in ***Hydra***, the largest of the 88 constellations, which sprawls a long way across the sky from its distinctively shaped 'head' of stars. A line from Regulus through ***Denebola*** (β Leonis) at the other end of the constellation of Leo points towards ***Arcturus*** (α Boötis), the brightest star in the northern hemisphere of the sky. The constellation of ***Boötes*** is sometimes described as 'kite-shaped' or 'shaped like the letter P'. Another line, down the 'back' of Leo, from ***Zosma*** (δ Leonis) through Denebola points in the general direction of ***Spica*** (α Virginis) the principal star in the constellation of ***Virgo***, a rough quadrilateral of moderately bright stars and fainter lines of stars extending outwards.

The two sides of the main quadrilateral, if extended, point eastwards towards the two principal stars, ***Zubeneschamali*** and ***Brachium*** (β and σ Librae, respectively), of the small zodiacal constellation of ***Libra***. The two small constellations of ***Corvus*** and ***Crater*** lie between Virgo and the long stretch of Hydra. A line from Arcturus through Spica, if extended far to the south across the sky, points to the distinctive constellation of ***Crux***.

The Winter Constellations

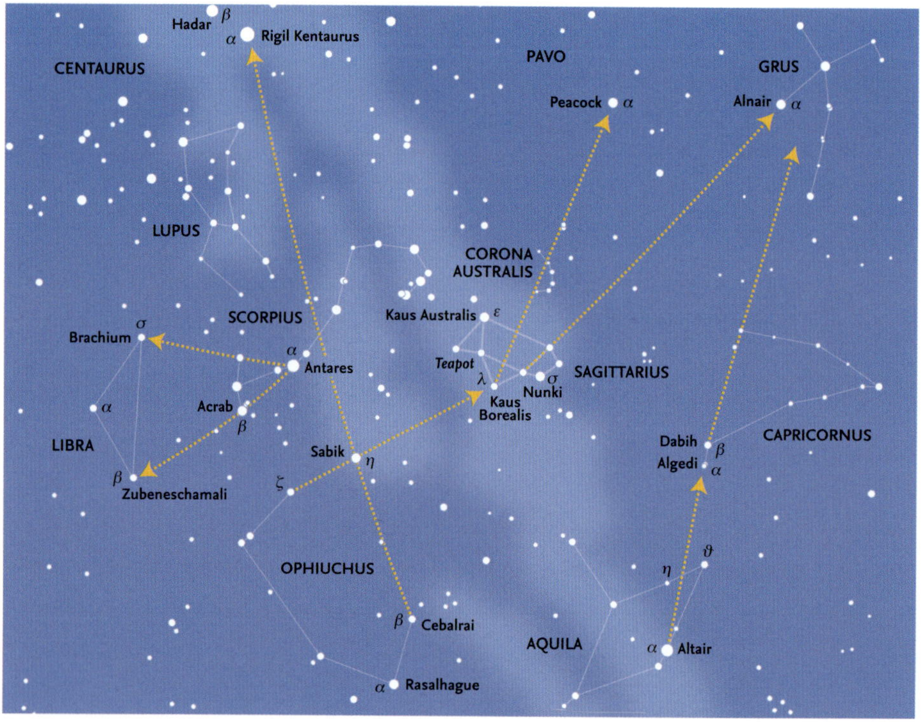

During the winter season, the Milky Way runs right across the sky, and the distinctive constellation of ***Scorpius*** is clearly visible, with bright red ***Antares*** (α Scorpii) and the constellation's 'sting' trailing behind it. The two sides of the 'fan' of stars immediately west of Antares, if extended, point to the two principal stars of ***Libra***, ***Zubeneschamali*** and ***Brachium*** (β and σ Librae, respectively).

North of Scorpius lies the large constellation of ***Ophiuchus***. The ecliptic runs through the southern portion of Ophiuchus, and the Sun spends far more time in that constellation than it does in neighbouring Scorpius. A line from ***Cebalrai*** (β Ophiuchi) in the north of the constellation, through ***Sabik*** (η Ophiuchi), if extended right across Scorpius, carries you to the two bright stars of ***Centaurus***, ***Rigil Kentaurus*** and ***Hadar*** (α and β Centauri, respectively). Between Libra and Centaurus lies the constellation of ***Lupus*** (the Wolf).

A line from the fainter star ζ Ophiuchi through ***Sabik*** points towards ***Kaus Borealis*** (λ Sagittarii) the 'lid' of 'the Teapot' of ***Sagittarius***. From there two lines radiate (on the west) across ***Corona Australis*** towards ***Peacock*** (α Pavonis) and (on the east) ***Alnair*** (α Gruis) in the small constellation of ***Grus***.

Also north of Scorpius is the constellation of ***Aquila***, with its brightest star, ***Altair*** (α Aquilae).

The diamond shape of the 'wings' of Aquila may be used to locate ***Algedi*** (α Capricorni), a prominent double star in ***Capricornus.*** A line extended through ***Dabih*** (β Capricorni) and the western side of Capricornus, also points in the direction of Grus.

Aquila consists of a diamond shape of stars, representing the body and wings of the eagle, together with a rather faint star, λ Aquilae, marking the 'head'.

Aquila (the Eagle) and the constellation of ***Cygnus*** (the Swan) represent birds 'flying' down the length of the Milky Way. This part of the Milky Way contains the ***Great Dark Rift***, an elongated dark region where the light from distant stars is obscured by intervening dust. The dark Rift is clearly visible even to the naked eye in a dark sky region.

The Spring Constellations

In spring, the constellation of ***Pegasus*** and 'the Great Square of Pegasus' are prominent in the north. However, the star at the northeastern corner, ***Alpheratz***, is actually α Andromedae, belonging to the adjacent constellation of ***Andromeda***. Between Andromeda and ***Cetus*** lie the two small constellations of ***Triangulum*** (the Triangle) and ***Aries*** (the Ram). A diagonal line across the Square from Alpheratz (α Andromedae) through ***Markab*** (α Pegasi) points to ***Sadalmelik*** (α Aquarii), just to the west of the 'Y-shaped' asterism of 'the Water Jar' in the zodiacal constellation of ***Aquarius***. Further extended, that line carries the observer to the centre of the triangular constellation of ***Capricornus.***

The constellation of ***Pisces*** consists of two lines of stars converging to the east of Pegasus. In mythological representations, the two fish are linked by ribbons, tied together at ***Alrescha*** (α Piscium). The distinctive asterism of 'the Circlet' lies at the western end of the southern line of stars. The western line of the Great Square passing from ***Scheat*** (β Pegasi) through Markab, crosses Aquarius and points to the bright, fairly isolated star, ***Fomalhaut*** (α Piscis Austrini) in the constellation of ***Piscis Austrinus*** (the Southern Fish) and then onward to ***Tiaki*** (β Gruis) in the centre of a cross of stars within ***Grus.*** The eastern line, from Alpheratz through ***Algenib*** (γ Pegasi), extended outwards indicates ***Diphda*** (β Ceti), which is actually the brightest star in Cetus and further across, ***Ankaa*** (α Phoenicis) in the constellation of ***Phoenix.***

The line between Markab and Algenib may be extended as an arc to lead to ***Menkar*** (α Ceti) in the 'tail' of Cetus. Diphda and Fomalhaut form the base of an almost perfect, large isosceles triangle, with ***Achernar*** (α Eridani) at the other apex. ***Eridanus*** is a long, trailing constellation that begins far to the north, near ***Rigel*** in ***Orion.*** West of Rigel lies the prominent cluster of the Pleiades, in the constellation of ***Taurus.***

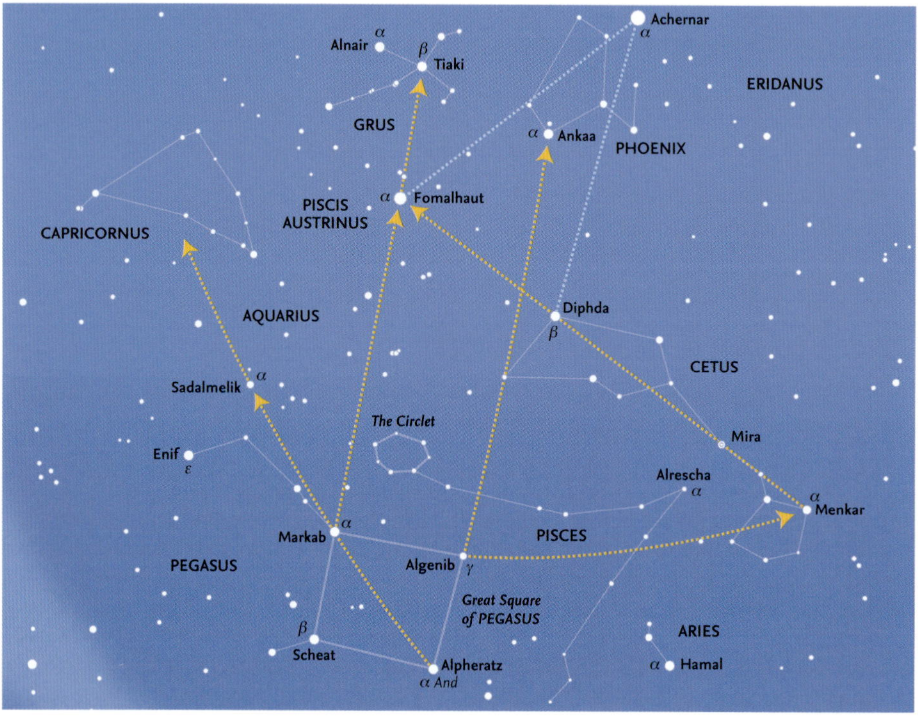

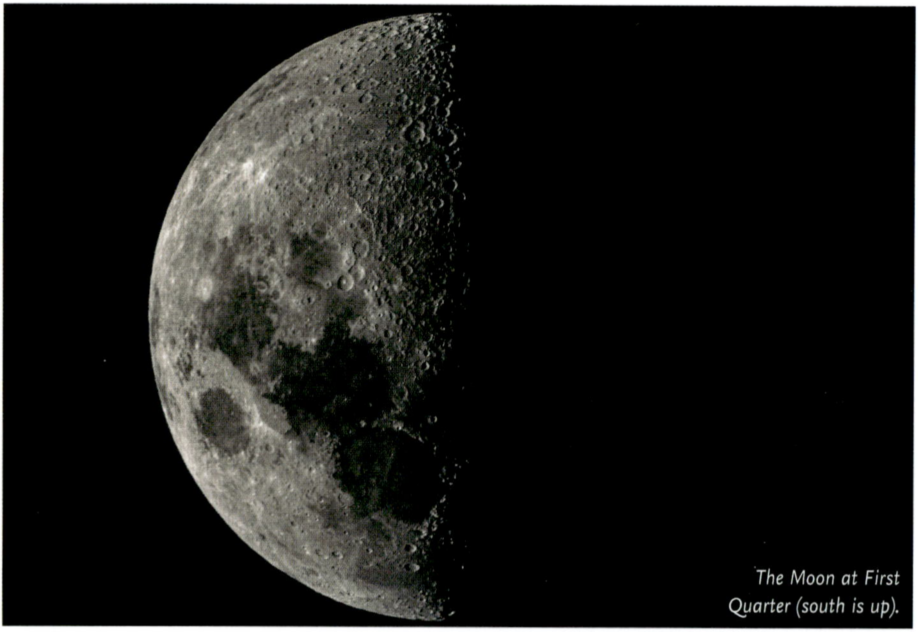

The Moon at First Quarter (south is up).

The Moon

The lunar phase cycle has a duration of 29.5 days and starts with the New Moon, when the near side of the Moon facing us is not illuminated by the Sun. The New Moon rises several hours after midnight and sets in the late afternoon or night; the First Quarter Moon rises in the late morning and sets around midnight; the Full Moon is above the horizon most of the night and the Last Quarter Moon rises around midnight and sets in the late morning.

Although the main features of the surface – the light highlands and the dark maria (seas) – may be seen with the naked eye, far more features may be detected with the use of binoculars or any telescope. The many craters are best seen when they are close to the ***terminator*** (the boundary between the illuminated and the non-illuminated areas of the surface), when the Sun rises or sets over any particular region of the Moon and the crater walls or central peaks cast strong shadows. These are best seen at crescent, quarter and gibbous phases. Most features become difficult to see at Full Moon, although this is the best time to see the bright ray systems surrounding certain craters. Accompanying the Moon map on the following pages is a list of prominent features, including the days in the lunar cycle when they are normally close to the terminator and thus easiest to see.

The dates of visibility vary slightly through the effects of ***libration***. Because the Moon's orbit is inclined to the Earth's equator and also because it moves in an ellipse, the Moon appears to rock slightly from side to side (and nod up and down). Over time a total of 59% of the lunar terrain can be observed. Areas near the ***limb*** (the edge of the Moon) may vary considerably in their location and visibility. This is easily noticeable with Mare Crisium (Sea of Crises, eastern limb of the Moon, in the western sky) and the craters Tycho (southern part of the Moon) and Plato (northern part). Another effect is that at crescent phases before and after New Moon, the normally non-illuminated portion of the Moon receives a certain amount of light, reflected from the Earth. This ***Earthshine*** may enable certain bright features (such as the craters Aristarchus, Kepler and Copernicus on the western limb of the Moon, in the eastern sky) to be detected.

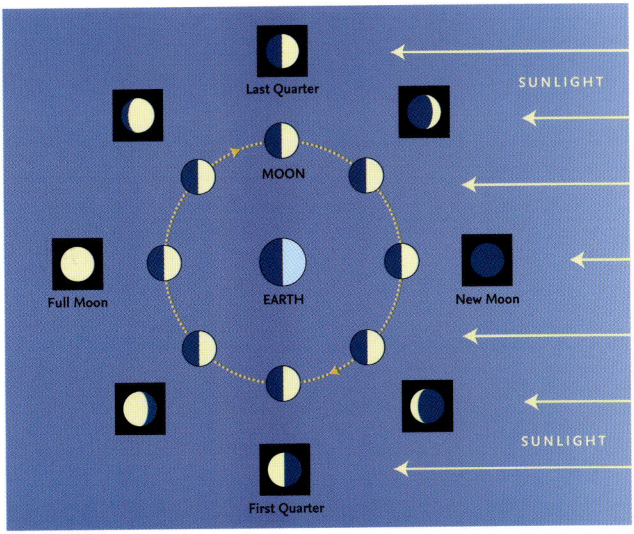

The Moon phases. *During its orbit around the Earth we see different portions of the illuminated side of the Moon's surface.*

Map of the Moon

Abulfeda	6:20
Agrippa	7:21
Albategnius	7:21
Aliacensis	7:21
Alphonsus	8:22
Anaxagoras	9:23
Anaximenes	11:25
Archimedes	8:22
Aristarchus	11:25
Aristillus	7:21
Aristoteles	6:20
Arzachel	8:22
Atlas	4:18
Autolycus	7:21
Barrow	7:21
Billy	12:26
Birt	8:22
Blancanus	9:23
Bullialdus	9:23
Bürg	5:19
Campanus	10:24
Cassini	7:21
Catharina	6:20
Clavius	9:23
Cleomedes	3:17
Copernicus	9:23
Cyrillus	6:20
Delambre	6:20
Deslandres	8:22
Endymion	3:17
Eratosthenes	8:22
Eudoxus	6:20
Fra Mauro	9:23
Fracastorius	5:19
Franklin	4:18
Gassendi	11:25
Geminus	3:17
Goclenius	4:18
Grimaldi	13-14:27-28
Gutenberg	5:19
Hercules	5:19
Herodotus	11:25
Hipparchus	7:21
Hommel	5:19
Humboldt	3:15
Janssen	4:18
Julius Caesar	6:20
Kepler	10:24
Landsberg	10:24
Langrenus	3:17
Letronne	11:25
Linné	6
Longomontanus	9:23

The numbers indicate the day or days in the lunar cycle when features are usually best visible.

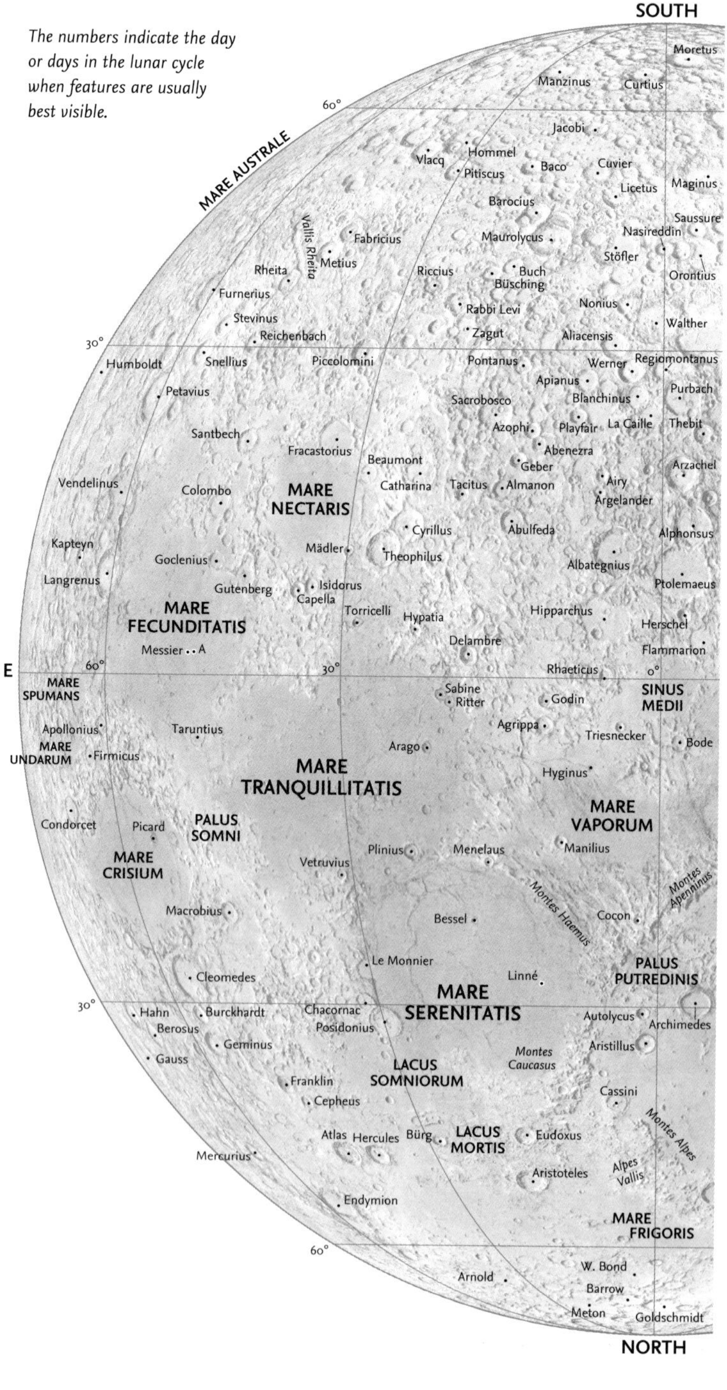

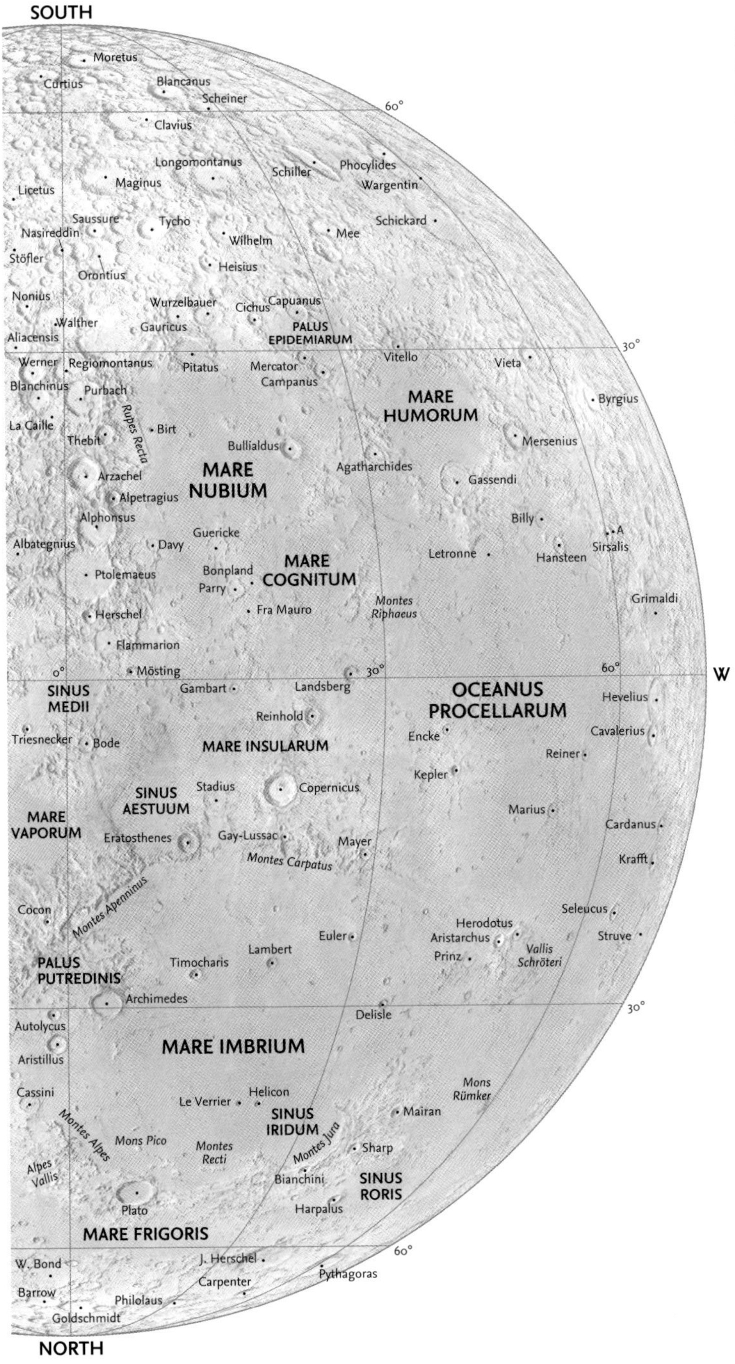
SOUTH
Moretus
Curtius
Blancanus
Scheiner
60°
Clavius
Longomontanus
Maginus
Schiller
Phocylides
Wargentin
Licetus
Saussure
Tycho
Nasireddin
Wilhelm
Mee
Schickard
Stöfler
Orontius
Heisius
Nonius
Wurzelbauer
Cichus
Capuanus
Walther
Gauricus
PALUS EPIDEMIARUM
Aliacensis
30°
Werner
Regiomontanus
Pitatus
Mercator
Campanus
Vitello
Vieta
Blanchinus
Purbach
MARE HUMORUM
Byrgius
Rupes Recta
La Caille
Birt
Thebit
Mersenius
Bullialdus
Agatharchides
Arzachel
MARE NUBIUM
Gassendi
Alpetragius
Alphonsus
Billy
Guericke
A
Albategnius
Davy
Sirsalis
Letronne
Hansteen
MARE COGNITUM
Ptolemaeus
Bonpland
Parry
Grimaldi
Fra Mauro
Montes Riphaeus
Herschel
Flammarion
0°
Mösting
30°
60°
W
SINUS MEDII
Gambart
Landsberg
OCEANUS PROCELLARUM
Hevelius
Reinhold
Triesnecker
Bode
MARE INSULARUM
Encke
Cavalerius
Reiner
Kepler
SINUS AESTUUM
Stadius
Copernicus
MARE VAPORUM
Marius
Eratosthenes
Gay-Lussac
Cardanus
Mayer
Montes Carpatus
Krafft
Montes Apenninus
Cocon
Seleucus
Herodotus
Aristarchus
Euler
Struve
Prinz
Vallis Schröteri
Lambert
PALUS PUTREDINIS
Timocharis
Archimedes
30°
Delisle
Autolycus
MARE IMBRIUM
Aristillus
Mons Rümker
Cassini
Helicon
Le Verrier
Mairan
SINUS IRIDUM
Montes Alpes
Montes Jura
Mons Pico
Montes Recti
Sharp
Alpes Vallis
Bianchini
SINUS RORIS
Plato
Harpalus
MARE FRIGORIS
60°
W. Bond
J. Herschel
Pythagoras
Carpenter
Barrow
Philolaus
Goldschmidt
NORTH

Eclipses in 2027

Lunar eclipses

There are two lunar eclipses in 2027 and both are penumbral. The Moon will darken slightly during the eclipse as it passes through the lighter peripheral part of Earth's shadow; the dramatic blood red Moon effect will not be seen as it only occurs during a total umbral lunar eclipse. The first penumbral eclipse takes place on 20 February (UTC), and it will be visible from Europe, Asia, western Australia, Africa, eastern regions of North America and South America. The maximum occurs at 23:12. The second penumbral eclipse takes place on 17 August, with the maximum occurring at 7:13. This eclipse will be visible from the western tip of Europe, northeast parts of Asia, eastern regions of Australia, New Zealand, northwest Africa, North America and South America.

Solar eclipses

There are two solar eclipses in 2027. The first is an annular eclipse on 6 February (UTC) with the maximum occurring at 15:59. The eclipse is visible from Chile and Argentina and it will appear as a partial eclipse from west and South Africa, Uruguay and Brazil. The duration of the annular eclipse is 7 minutes 51 seconds. There will be a total eclipse on 2 August visible from Spain, Morocco, Algeria, Libya, Egypt, Saudi Arabia, Yemen and Somalia. Maximum occurs at 10:06, the duration is 6 minutes 23 seconds. A partial solar eclipse will be visible from Europe, Africa, the Middle East, west and south Asia.

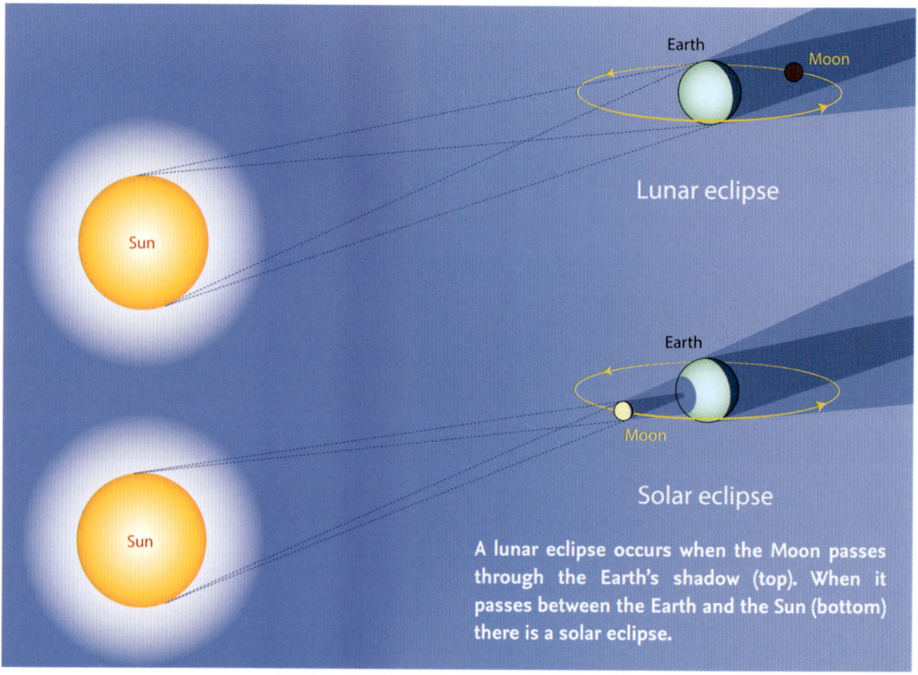

A lunar eclipse occurs when the Moon passes through the Earth's shadow (top). When it passes between the Earth and the Sun (bottom) there is a solar eclipse.

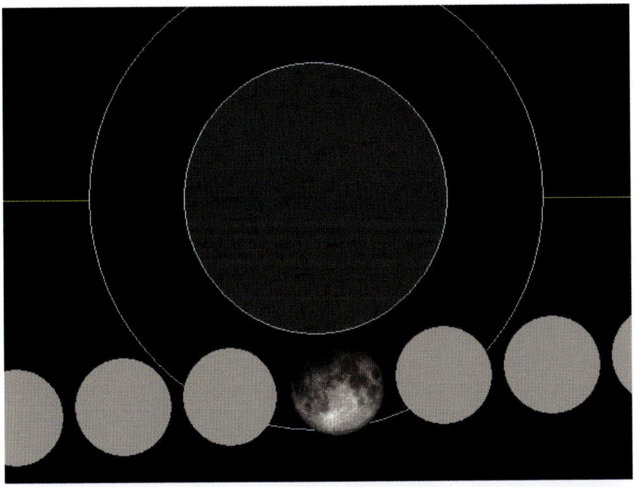

The path of the Moon as it passes through the penumbral shadow of the Earth on 20 February. Maximum eclipse occurs at 23:12 UT.

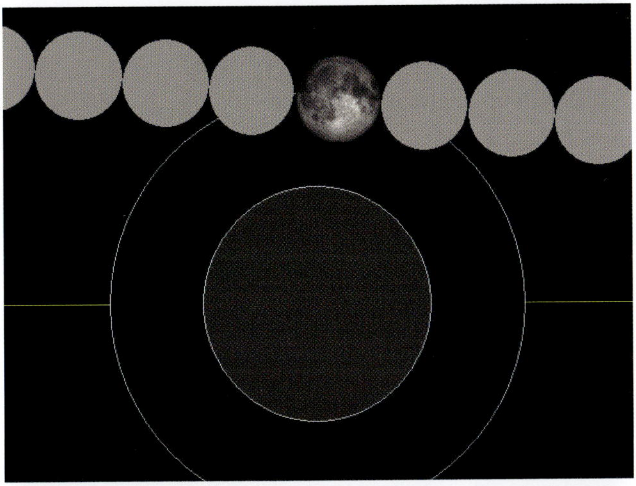

The path of the Moon as it passes through the penumbral shadow of the Earth on 17 August. Maximum eclipse occurs at 07:13 UT.

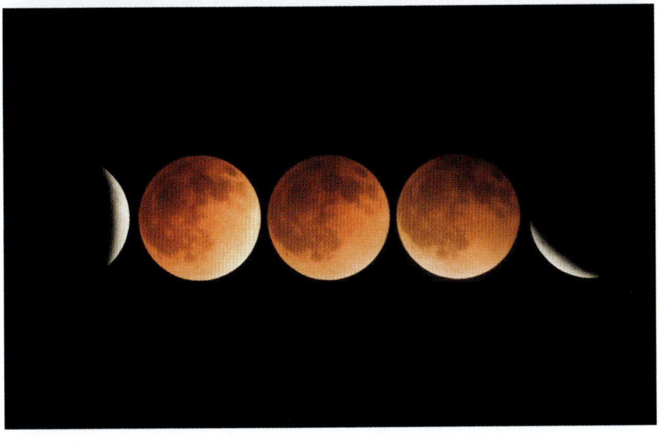

A sequence of five images showing the central stages of a lunar eclipse, taken from Cascade-Siskiyou National Monument, Oregon in 2014.

The Planets in 2027

Mercury and Venus

Mercury reaches greatest eastern elongation three times during 2027 – it will be seen in the early evening of 3 February at mag. -0.5, 28 May (mag. 0.4) and 24 September (mag. 0.1, setting with Venus and Mars). Mercury switches to an appearance at dawn when it is at greatest western elongation on 17 March at mag. 0.2 (with Venus further south), 16 July at mag. 0.4 (with Venus closer to the Sun) and 4 November at mag. -0.5 (Jupiter sitting higher to the south). Its magnitude oscillates throughout the year – it will be brightest in April (-2.4) and faintest in October (6.3).

Venus starts the year leading the Sun in the dawn sky. It reaches greatest western elongation on 3 January appearing as the 'morning star' at mag. -4.5. Venus dims gradually throughout the year to mag. -3.9, at its faintest from May. Venus will appear in the evening sky beyond 11 August, remaining at around mag. -3.9 for the rest of the year.

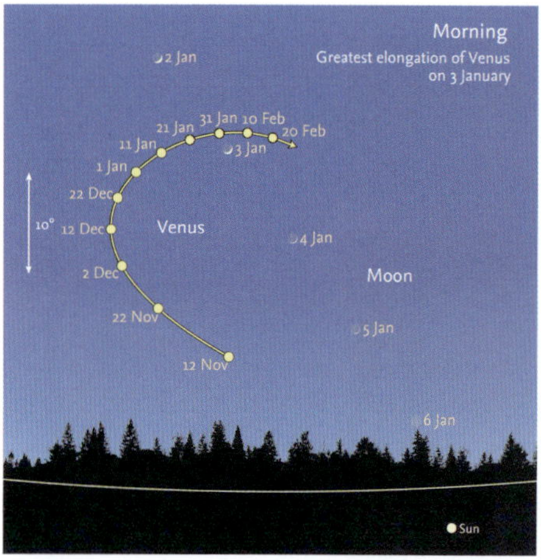

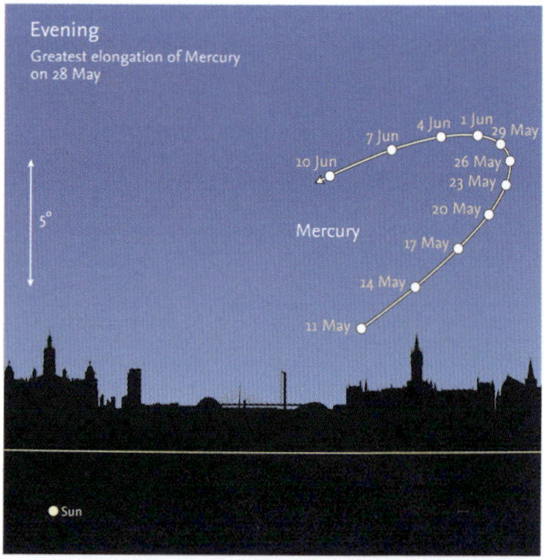

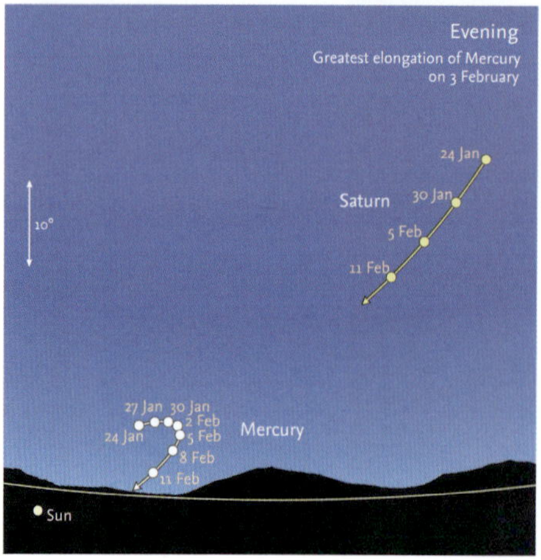

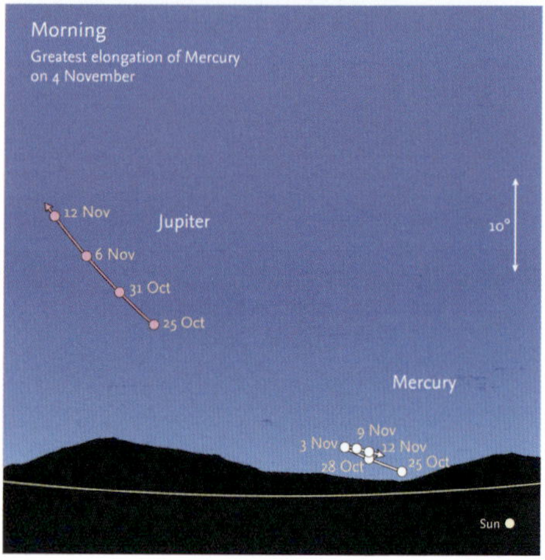

Mars

Mars starts the year along the ecliptic in ***Leo***, rising late in the evening. It enters retrograde (westward) motion on 10 January as the Earth overtakes it in its orbit around the Sun, reverting to direct (eastward) motion on 1 April. Mars moves into ***Virgo*** in July and then progresses into ***Libra*** in September. It shifts into ***Scorpio*** and then ***Ophiuchus*** in October, after which it settles into ***Sagittarius*** in November and eventually sets with the Sun in December.

Mars brightens to its peak magnitude of -1.3 in February while it sits in Leo. Its apparent size in the sky will increase as it reaches opposition on 19 February 2027. It gradually dims to its faintest in October (mag. 1.4) and it will brighten again towards the end of the year. As it takes longer to complete its orbit than the Earth, Mars does not come to opposition every year. Oppositions occur during a period of retrograde motion, when the planet appears to move westwards against the pattern of distant stars. Because of its eccentric orbit, which carries it at differing distances from the Sun (and Earth), not all oppositions of Mars are equally favourable for observation. The relative positions of Mars and the Earth are shown here. It can be seen that the opposition of 2018 was very close and thus favourable for observation. By comparison, opposition in 2027 will be at a far greater distance, so the planet will not appear as large as it did in 2018.

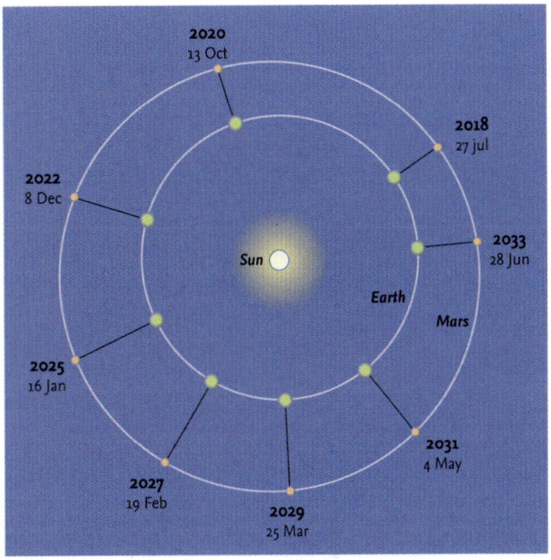

The oppositions of Mars between 2018 and 2033. The next opposition takes place on 19 February 2027.

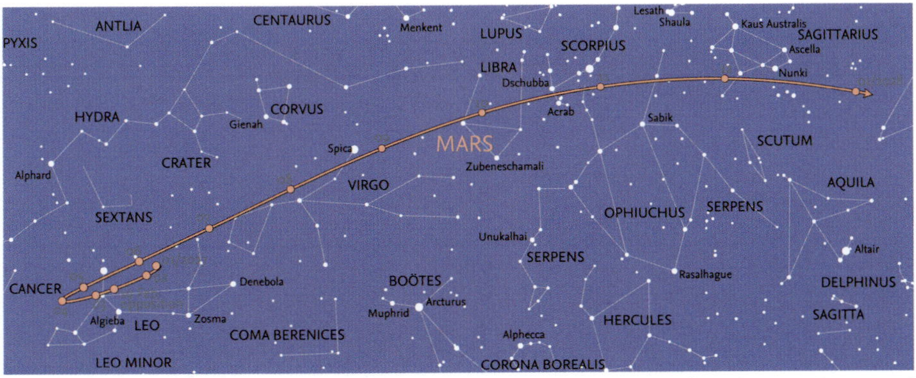

The path of Mars in 2027. Mars reaches opposition on 19 February. It enters retrograde motion on 10 January and reverts back to direct motion on 1 April. Mars is visible in the evening for most of the year.

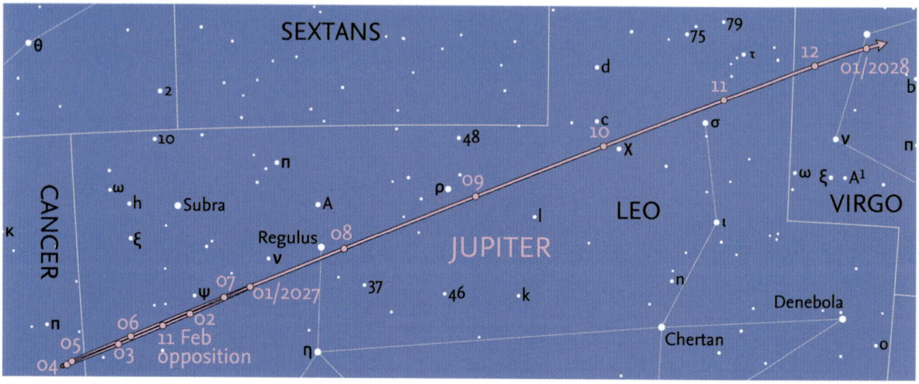

The path of Jupiter in 2027. Jupiter reaches opposition on 11 February. It ends retrograde motion on 13 April and re-enters retrograde westward motion in January 2028.

Jupiter and Saturn

Jupiter starts its journey in 2027 moving westward in retrograde motion along the ecliptic in ***Leo***. It reaches opposition on 11 February at peak brightness, mag. -2.6. Jupiter enters ***Cancer*** in March and ends retrograde motion on 13 April, staying in direct motion until January 2028. It moves back into Leo in May and sets with the Sun in late August, lost in the glare of the Sun and dimming to mag. -1.7. In October Jupiter reappears in the early morning sky and enters ***Virgo*** in November.

Jupiter's four largest satellites are readily visible in binoculars. Not all four are visible all the time, they are sometimes hidden behind the planet or invisible in front of it. ***Io***, the closest to Jupiter, orbits in just under 1.8 days, and ***Callisto***, the farthest away, takes about 16.7 days. In between are ***Europa*** (c. 3.6 days) and the largest, Ganymede (c. 7.2 days).

Saturn is in ***Cetus*** at the start of the year, it moves into ***Pisces*** in late February, visible in the early evening at its faintest (mag. 0.8). The ringed planet swings into the dawn sky in May and gradually pulls away from the Sun, brightening to its peak at opposition on 18 October (mag. 0.1). This will be a good time to look at the planet, its rings will be inclined by 13° to our line of sight (allowing observers to see more of the southern side). It enters retrograde motion on 9 August and will be visible above the horizon before midnight from September. Saturn ends its retrograde motion in Pisces on 24 December.

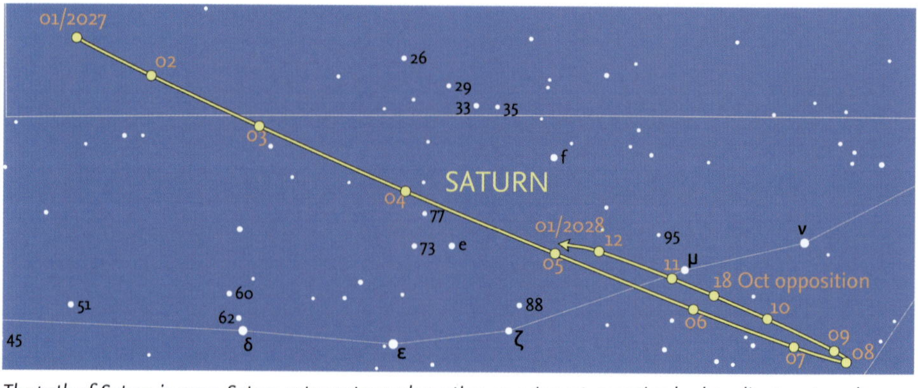

The path of Saturn in 2027. Saturn enters retrograde motion on 9 August, reverting back to direct eastward motion on 24 December. It reaches opposition on 18 October.

Uranus and Neptune

Uranus sits in ***Taurus*** for the whole of 2027. It leads the Sun in the dawn sky in June–July. It ends its retrograde motion on 8 February and re-enters retrograde on 15 September, moving westwards until 12 February 2028. Uranus will reach opposition on 30 November (mag. 5.6).

Neptune moved into ***Pisces*** in May 2022; it will stay in Pisces until August 2039. Neptune appears in the dawn sky from April, returning to the evening sky from September. It enters retrograde motion on 10 July at mag. 7.8. It will reach opposition on 28 September (mag. 7.7). Neptune reverts to direct motion on 15 December, fainter at mag. 7.8 and will continue moving eastwards until 11 July 2028.

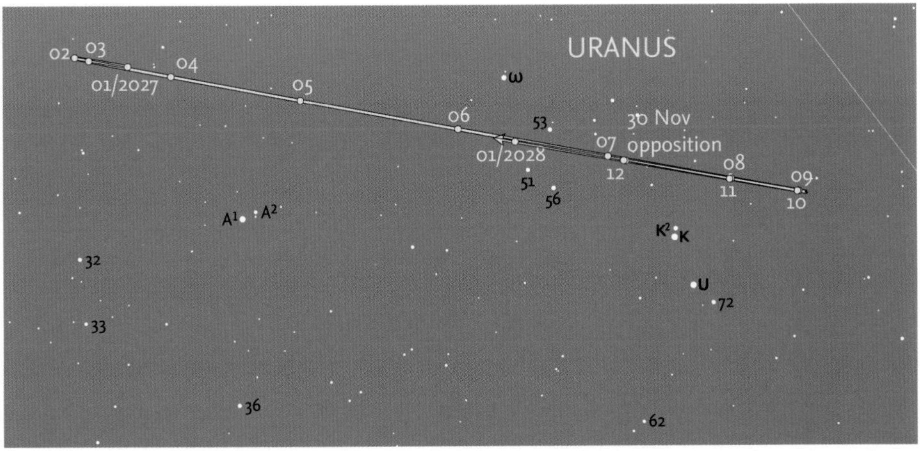

The path of Uranus in 2027 in Taurus. Uranus ends retrograde motion on 8 February, re-entering on 15 September. It reaches opposition on 30 November.

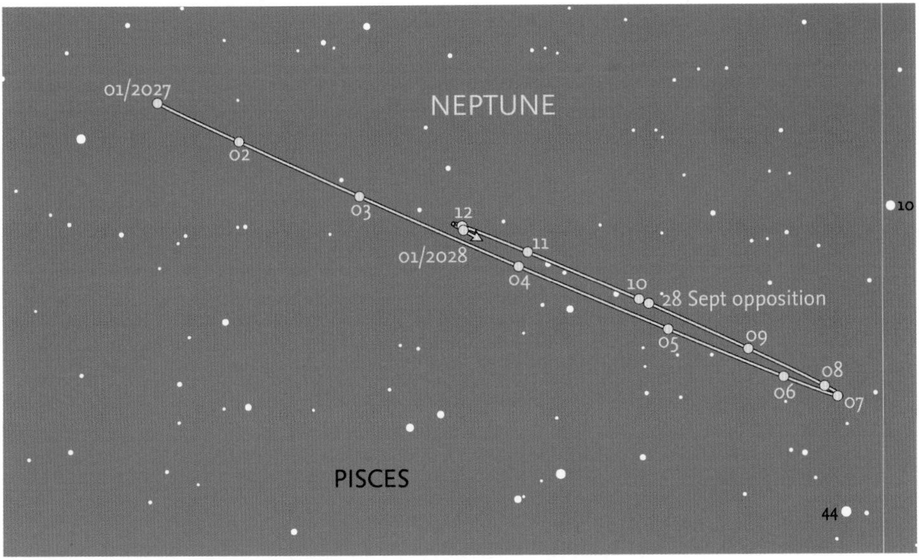

The path of Neptune in 2027 in Pisces. Neptune enters retrograde motion on 10 July, reaching opposition on 28 September. It reverts to direct eastward motion on 15 December.

Minor Planets in 2027

Three minor planets come to opposition and rise above magnitude 9.0 in 2027. They are closest to Earth at this point and reach their highest point in the sky on the meridian around midnight – this is the best time to see them. The year starts with the brightening of the dwarf planet ***(1) Ceres***, reaching a peak magnitude of 6.8 at opposition on 7 January 2027 when it will be at a distance of 1.63 AU (Astronomical Units, equivalent to the average distance from the Earth to the Sun). Named after the Roman goddess of agriculture, Ceres is the only confirmed dwarf planet that lies within the orbit of Neptune, and indeed orbits in the asteroid belt, a vast ring of rocks between Mars and Jupiter, making up 25% of the total mass of said region. Ceres completes its orbit of the Sun in 4.6 years.

On 9 February the asteroid ***(27) Euterpe*** will lie in ***Leo***, reaching a peak magnitude of 8.9 at opposition, at a distance of 1.06 AU. On 21 September ***(29) Amphitrite*** will reach magnitude 8.8 in ***Pisces*** at opposition, passing within 1.45 AU of Earth.

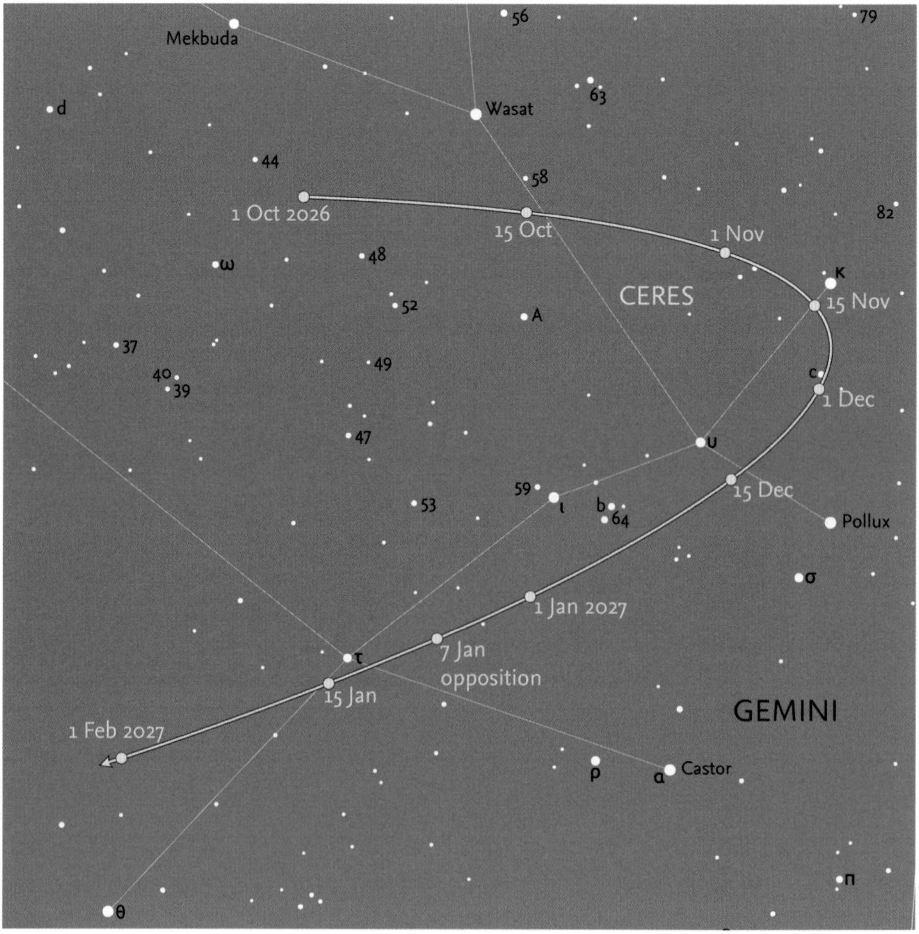

The path of (1) Ceres in 2027. (1) Ceres reaches opposition on 7 January, at magnitude 6.8.

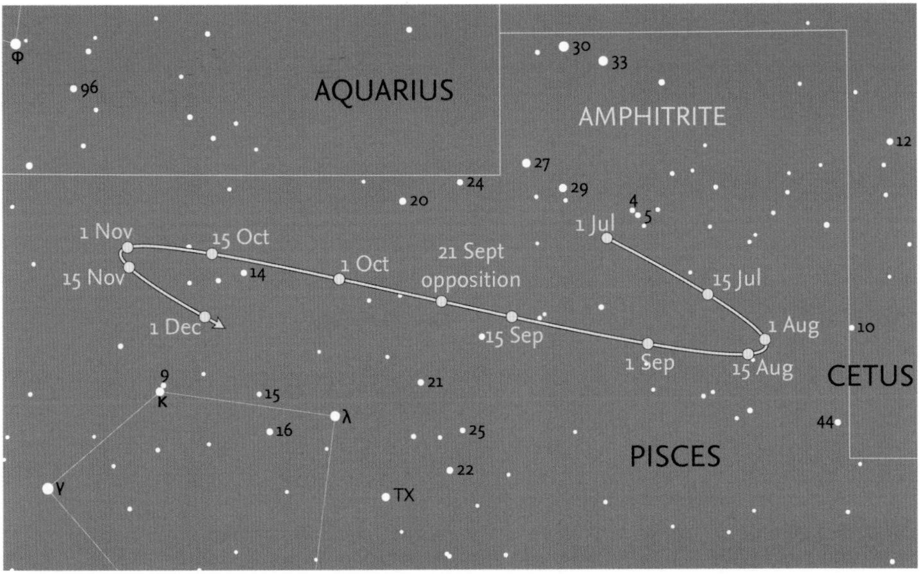

The path of (29) Amphitrite in 2027. (29) Amphitrite reaches opposition on 21 September, at magnitude 8.8.

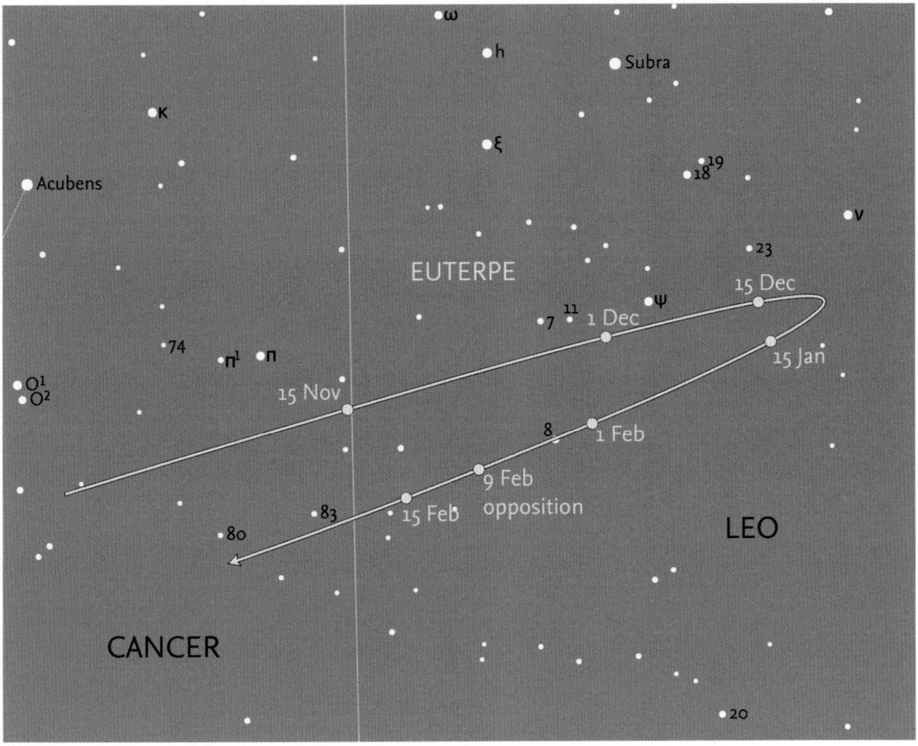

The path of (27) Euterpe in 2027. (27) Euterpe reaches opposition on 9 February, at magnitude 8.9.

Comets in 2027

Although comets may occasionally become very striking objects in the sky, their occurrence and particularly the existence or length of any tail and their overall magnitude are notoriously difficult to predict. Naturally, it is only possible to predict the return of periodic comets (whose names have the prefix 'P'). Many comets appear unexpectedly (these have names with the prefix 'C'). Bright, readily visible comets such as C/1995 Y1 Hyakutake, C/1995 O1 Hale-Bopp, C/2006 P1 McNaught or C/2020 F3 NEOWISE are rare. Most periodic comets are faint and only a very small number ever become bright enough to be easily visible with the naked eye or with binoculars.

Comet 2P/Encke will be visible early in the evening in January, moving through the constellations of Aquarius and Capricornus in the early part of the year and brightening to magnitude 5. It will make its closest approach to Earth (perigee) on 3 February 2027, passing within a distance of 0.62 AU. Comet 2P/Encke has a period of 3.3 years, the trail of debris left in its wake is the source of the Northern and Southern Taurid meteor showers, which occur in late October and November.

Comet 45P/Honda-Mrkos-Pajdusakova will progress through Orion, Taurus, Gemini, Cancer and Leo in August and September, brightening to around magnitude 8.5 as it approaches perihelion on 31 August. It will be very low in the sky just before sunrise. Also, a short-period comet, it completes its orbit of the Sun in 5.3 years.

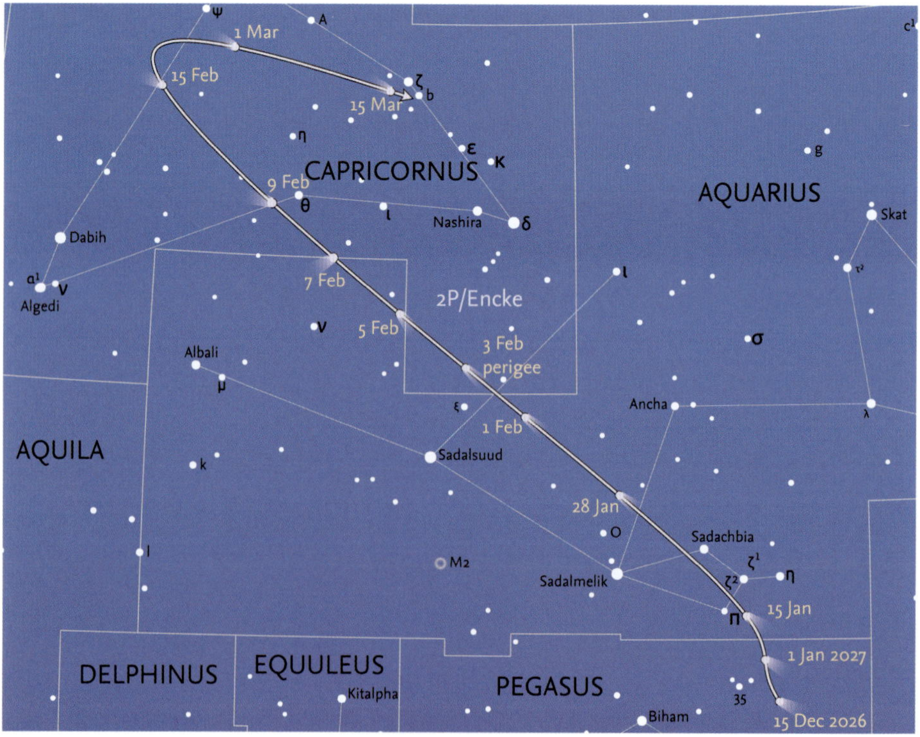

Comet 2P/Encke reaches magnitude 5 at its closest approach to Earth on 3 February.

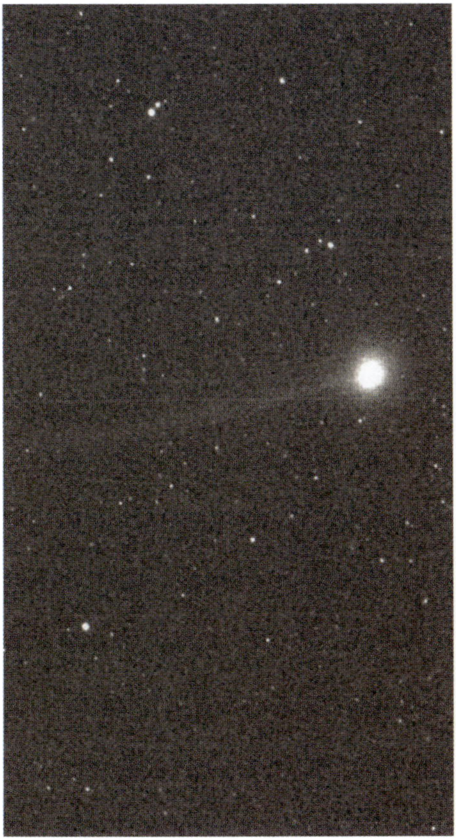

Comet 2P/Encke.

Comet 45P/Honda-Mrkos-Pajdusakova.

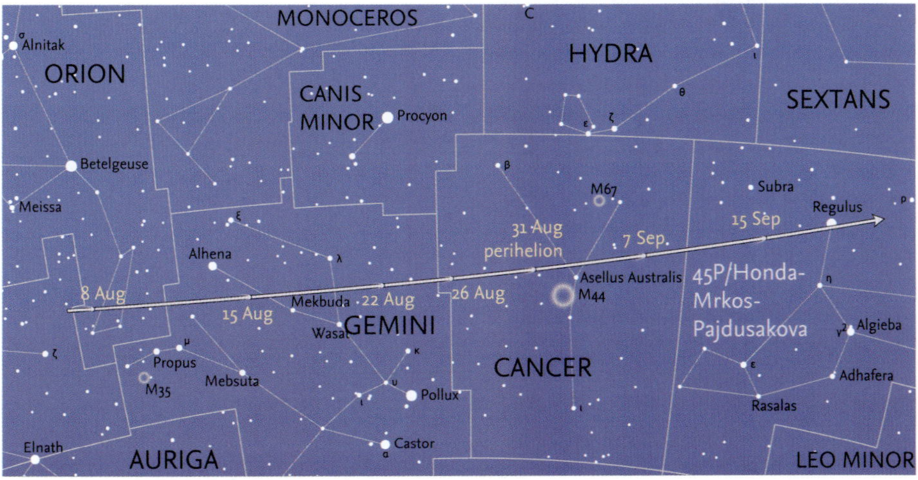

Comet 45P/Honda-Mrkos-Pajdusakova reaches magnitude 8.5 as it approaches perihelion on 31 August.

Introduction to the Month-by-Month Guide

The monthly charts

The pages devoted to each month contain a pair of charts showing the appearance of the night sky, looking north and looking south. The charts are drawn for the latitude of 35°S, so observers farther south will see slightly more of the sky on the southern horizon, and slightly less on the northern. Stars close to the horizon are always dimmed by atmospheric absorption (this is called extinction), so sometimes the faintest stars marked on the charts may not be visible.

The charts are drawn to show the appearance at 23:00 local time in Sydney, Australia for the 1st of each month. The same appearance will apply an hour earlier (22:00) on the 15th, and yet another hour earlier (21:00) at the end of the month (shown as the 1st of the following month). In New Zealand, the previous period of DST ends on the first Sunday of April (4 April 2027) and starts on the last Sunday of September (26 September 2027). In those Australian states with DST, it ends on the first Sunday in April (4 April 2027) and starts on the first Sunday in October (3 October 2027). Daylight Saving Time (DST) is not used in South Africa. The appropriate times are shown on the monthly charts. Times of specific events in the diagrams are shown in local time for Sydney.

The charts may be used for earlier or later times during the night. To observe two hours earlier, use the charts for the preceding month; for two hours later, the charts for the next month.

Tips for observing the night sky

A successful observing session requires planning, flexibility and patience. Check the weather forecast, sunset and moonrise and moonset times and choose a location where you have a clear view of the horizon wherever possible. Bring your star guide or app on a mobile device, a red torch to aid your night vision and a pair of binoculars and a tripod to keep them steady. You may be outside for a while – it's advised that you wear warm clothes, socks and shoes and bring hot drinks if it's a cold night.

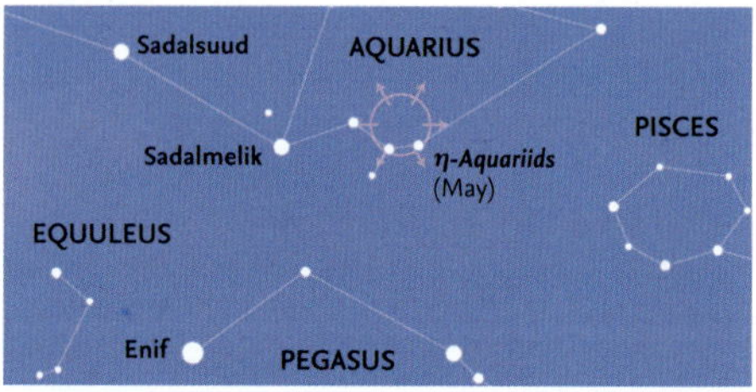

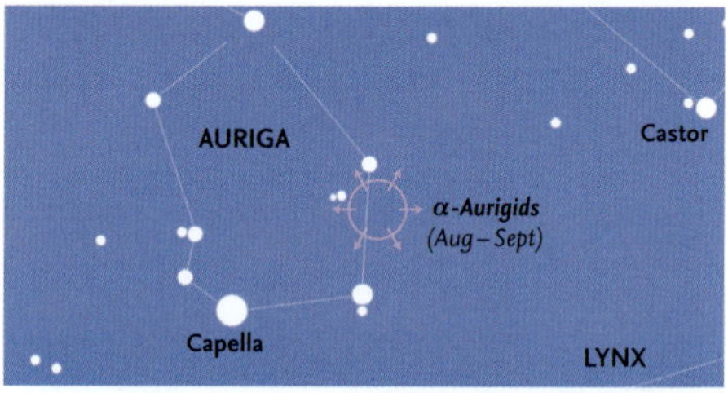

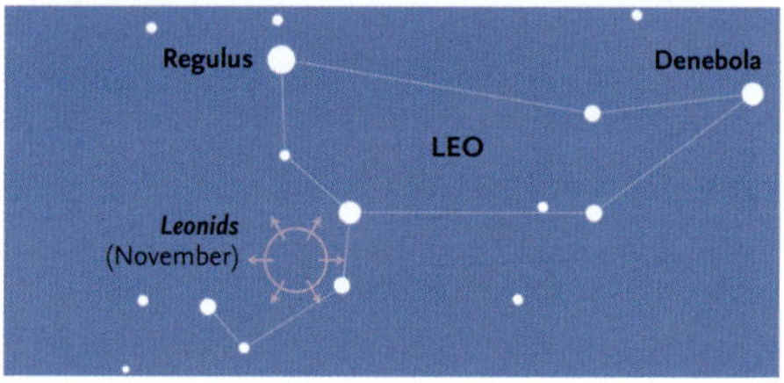

If you are planning on looking for meteors or looking up at the night sky for a while you may wish to bring a deckchair or picnic blanket with you. When looking at faint objects (unlike the Moon) you will need to give your eyes 10–20 minutes to adapt to the dark and achieve night vision. If you are using your phone, switch it to night vision mode and use a red torch to find your way around so that you maintain your night vision.

Astrophotography

You can capture the Moon and five planets visible to the naked eye or the brightest stars using your smartphone. The challenge is keeping your phone steady and gaining control over the focus and exposure (if the camera allows it). It is advised to attach the phone to a tripod, or you could use an adaptor to attach your phone to the eyepiece of a telescope – this will allow

Shower	Dates of activity 2027	Date of maximum 2027	Possible hourly rate
α-Centaurids	January 28–February 21	February 8	6
γ-Normids	February 25–March 28	March 14	6
π-Puppids	April 15–April 28	April 24	var.
η-Aquariids	April 19–May 28	May 5–6	40
α-Capricornids	July 3–August 15	July 30	5
Southern δ-Aquariids	July 12–August 23	July 30	25
Piscis Austrinids	July 15–August 10	July 29	5
Perseids	July 17–August 24	August 12–13	150
α-Aurigids	August 28–September 5	September 1	6
Orionids	October 2–November 7	October 21–22	15
Southern Taurids	September 10–November 20	November 5	5
Northern Taurids	October 20–December 10	November 12–13	5
Leonids	November 6–November 30	November 17–18	15
Phoenicids	November 28–December 9	December 2	var.
Puppid-Velids	December 1–December 15	December 7	10
Geminids	December 4–December 17	December 14	120

you to align it correctly with minimal effort. You can use the zoom function on your camera to change the magnification of the image.

There are many apps available that will provide more control of your camera settings and allow you to take better quality images: NightCap Pro, ProCam and AstroShader for iOS and Open Camera, DeepSkyCamera or Camera FV-5 for Android. For the Moon and bright planets, your camera should auto lock and focus, otherwise manually set the focus to infinity and keep the exposure short (a fraction of a second if it is a gibbous or Full Moon, one or two seconds for a thin crescent Moon and planets). You could control the shutter remotely via the volume control on wireless headphones, or set a timer on your phone (3 seconds delay). For fainter objects, increase the exposure and the ISO (the sensitivity of the camera detector); however, a higher ISO setting will introduce more noise. Trial and error will help you choose the best settings. Some apps allow images to be stacked; this means you can overlay images to increase brightness further. This is useful for the constellations, clusters such as the Pleiades and nebulous objects such as the Orion Nebula and Andromeda.

Meteors

Details of specific meteor showers are given in the months when they come to maximum. Radiants that lie below the horizon or in constellations not readily visible in that month are not marked. For this reason, special charts for the Eta (η) Aquariids (May), the Alpha (α) Aurigids (August and September) and the Leonids (November) are given here. Meteors from such showers may still be seen, because the most effective region for seeing meteors is some 40–45° from the radiant, and that area of sky may well be above the horizon. A table of the best meteor showers visible during the year is also given here. The rates given are those that an experienced observer might see under ideal conditions. Generally, the observed rates will be far lower.

The photographs

One or more photographs of constellations visible in certain specific months are included. It should be noted, however, that several factors affect observation: differences between the sensitivity of different individuals to faint starlight including colour, the degree to which they have become adapted to the dark, the local altitude of the star and the atmospheric conditions. Also, the exposure time can be altered on a digital camera, thus enhancing the intensity and colour of the object. Another factor to consider is the view of the object, telescopes invert the image and in some cases the image is mirror-reversed (left and right are swapped). Each telescope will have a different

A fireball (with flares approximately as bright as the Full Moon), photographed against a weak auroral display from Tarbat Ness by Denis Buczynski in Scotland on 22 January 2017.

optical configuration. Corrective optics can be used; however, the more optics are included the more the light is attenuated as it passes through to the eyepiece. Binoculars provide an image with the correct orientation.

The Moon calendar

The Moon calendar shows the phase of the Moon for every day of the month. Also shown is the ***age*** of the Moon (the day in the ***lunation***), beginning at New Moon, which may be used to determine the best time for observation of specific lunar features.

The Moon

The section on the Moon includes details of any lunar or solar eclipses that may occur during the month. Similar information is given about any important occultations. Mainly, however, this section summarizes when the Moon passes close to planets or the five prominent stars close to the ecliptic. The dates when the Moon is closest to the Earth (at ***perigee***) and farthest from it (at ***apogee***) are shown in the monthly calendars, and only the nearest and farthest points during the year are mentioned.

The planets and minor planets

Brief details are given of the location, movement and brightness of the planets from Mercury to Saturn throughout the month. Mercury and Venus are normally easiest to see around either eastern or western elongation, in the evening or morning sky, respectively.

The dates at which the superior planets reverse their motion (from direct motion to ***retrograde***, and retrograde to direct) and of opposition (when a planet generally reaches its maximum brightness) are given. Jupiter and Saturn are normally easiest to see around opposition, which occurs every year. Mars, by contrast, moves relatively rapidly against the background stars and in some years never comes to opposition. The distant outer planets, especially Saturn, Uranus and Neptune, may spend most or all of the year in a single constellation.

Uranus is normally magnitude 5.7 to 5.9, and thus at the limit of naked-eye visibility under exceptionally dark skies, but bright enough to be readily visible in binoculars. Similar considerations apply to Neptune, it is always fainter (generally magnitude 7.8 to 8.0); however, it is still visible in most binoculars. The charts on page 25 show their positions during 2027.

In any year, few minor planets ever become bright enough to be detectable in binoculars. Just one, (4) Vesta, on rare occasions brightens sufficiently for it to be visible to the naked eye, with no such occurrences in 2027. Our limit for visibility is magnitude 9.0 and details and charts are given for those objects that exceed that magnitude during the year, normally around opposition. Minor planet charts for 2027 are on pages 26 and 27.

The ecliptic charts

Although the ecliptic charts are primarily designed to show the positions and motions of the major planets, they also show the motion of the Sun during the month. The light-tinted area shows the area of the sky that is invisible during daylight, but the darker area gives an indication of which constellations are likely to be visible at some time of the night.

The monthly calendar

For each month, a calendar shows details of significant events, including when planets are close to one another in the sky, close to the Moon, or close to any one of five bright stars that are spaced along the ecliptic. The times shown are given in Universal Time (UT).

The diagrams of interesting events

Each month, a number of diagrams show the appearance of the sky when certain events take place, shown in local time for Sydney, Australia unless otherwise specified. However, the exact positions of celestial objects and their separations greatly depend on the observer's position on Earth. When the Moon is one of the objects involved, because it is relatively close to Earth, there may be very significant changes from one location to another. Close approaches between planets or between a planet and a star are less affected by changes of location, which may thus be ignored.

The diagrams showing the appearance of the sky are drawn for the latitude 35°S and longitude 150°E (approximately that of Canberra, Australia), so they will be approximately correct for much of Australia. However, for an observer farther north (say, in Brisbane or Darwin), a planet or star listed as being north of the Moon will appear even farther north, whereas one south of the Moon will appear closer to it – or may even be hidden (occulted) by it. For an observer at a latitude greater than 35°S (such as in New Zealand), there will be corresponding changes in the opposite direction: for a star or planet south of the Moon, the separation will increase, and for one north of the Moon, the separation will decrease. This is particularly important when occultations occur, which may be visible from one location, but not another.

The details given regarding the positions of the various bodies should be used as a guide to their location. A similar situation arises with the local times that are shown.

Data used in this Guide

The data given in this Guide, such as timings and distances between objects, have been determined by the following sources: the Astronomical Applications Department of the US Naval Observatory; the Sky Events Calendar by Fred Espenak and Sumit Dutta (NASA's Goddard Space Flight Center); the NASA Jet Propulsion Laboratory (JPL) Horizons System; the International Astronomical Union's (IAU) Minor Planet Center; In-The-Sky.org.

Key to the symbols used on the monthy star maps.

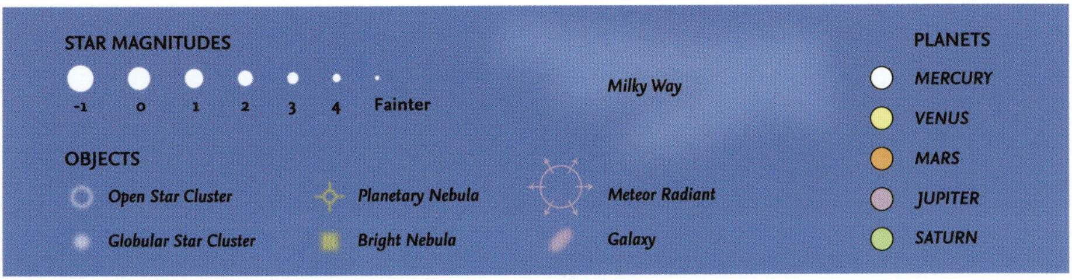

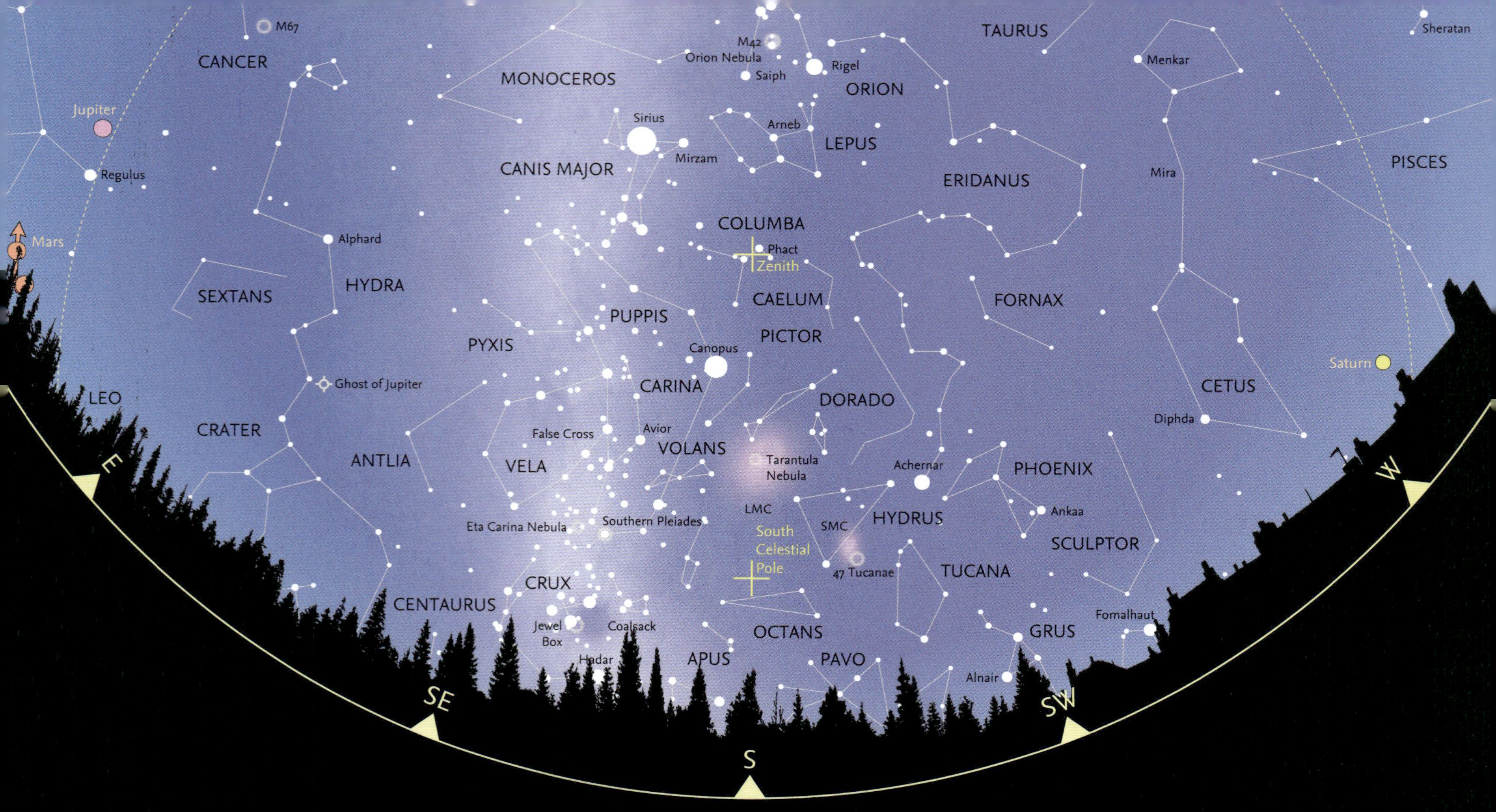

January – Looking South

1 January	23:00	(00:00 DST)
15 January	22:00	(23:00 DST)
1 February		

January – Looking South

Crux is quite prominent as it rises in the east, and even ***Rigil Kentaurus*** and ***Hadar*** (α and β Centauri), although low, are becoming easier to see. The whole of ***Carina*** is visible, with both the ***Eta Carinae Nebula*** and the ***Southern Pleiades*** between Crux and the ***False Cross*** of stars from the constellations of Carina and ***Vela***. ***Canopus*** (α Carinae) is high in the south, roughly three-quarters of the way from the horizon to the zenith. Also in the south, but slightly lower, is the ***Large Magellanic Cloud*** (LMC), a satellite dwarf galaxy 160,000 light years away, consisting of an estimated 20 billion stars. ***Achernar*** (α Eridani) is prominent between the constellations of ***Hydrus*** and ***Phoenix***. The ***Small Magellanic Cloud*** (SMC), another satellite galaxy 200,000 light years away and ***47 Tucanae***, a globular cluster of around 10,000 stars are well-placed for observation alongside Hydrus. ***Fomalhaut*** (α Piscis Austrini) may be glimpsed low on the horizon towards the west, as may ***Peacock*** (α Pavonis) farther towards the south. The whole of ***Pavo*** and ***Grus*** is visible, together with most of ***Indus***.

Meteors

There is one minor shower at the start of the year for observers in the southern hemisphere, the ***α-Centaurids*** (active late January to February), which has a low maximum rate of around six per hour.

The constellation Eridanus, the river.

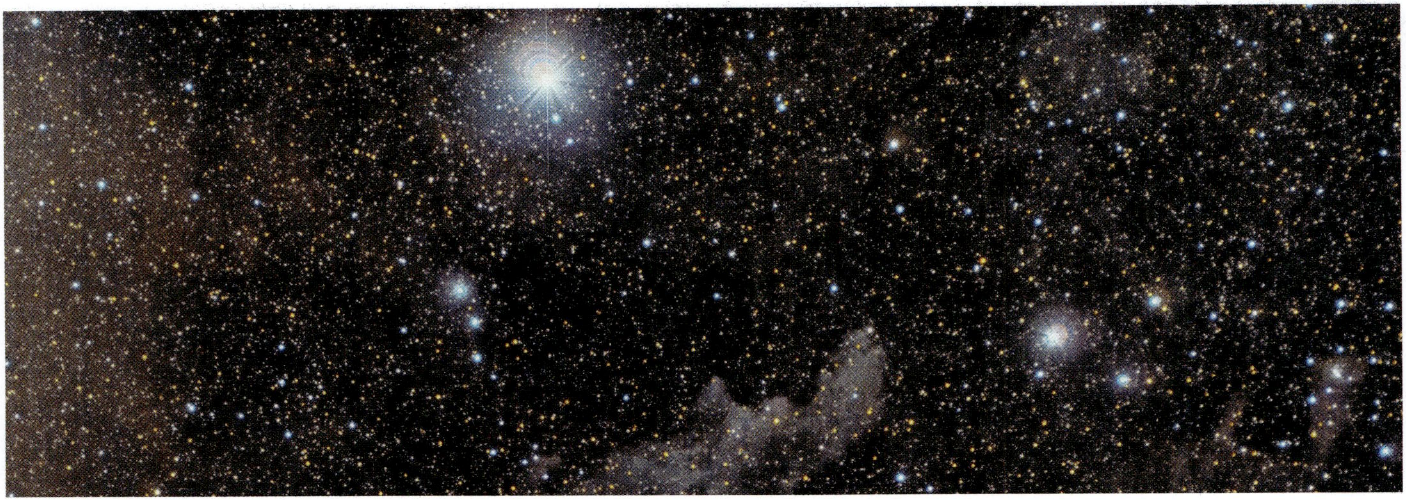

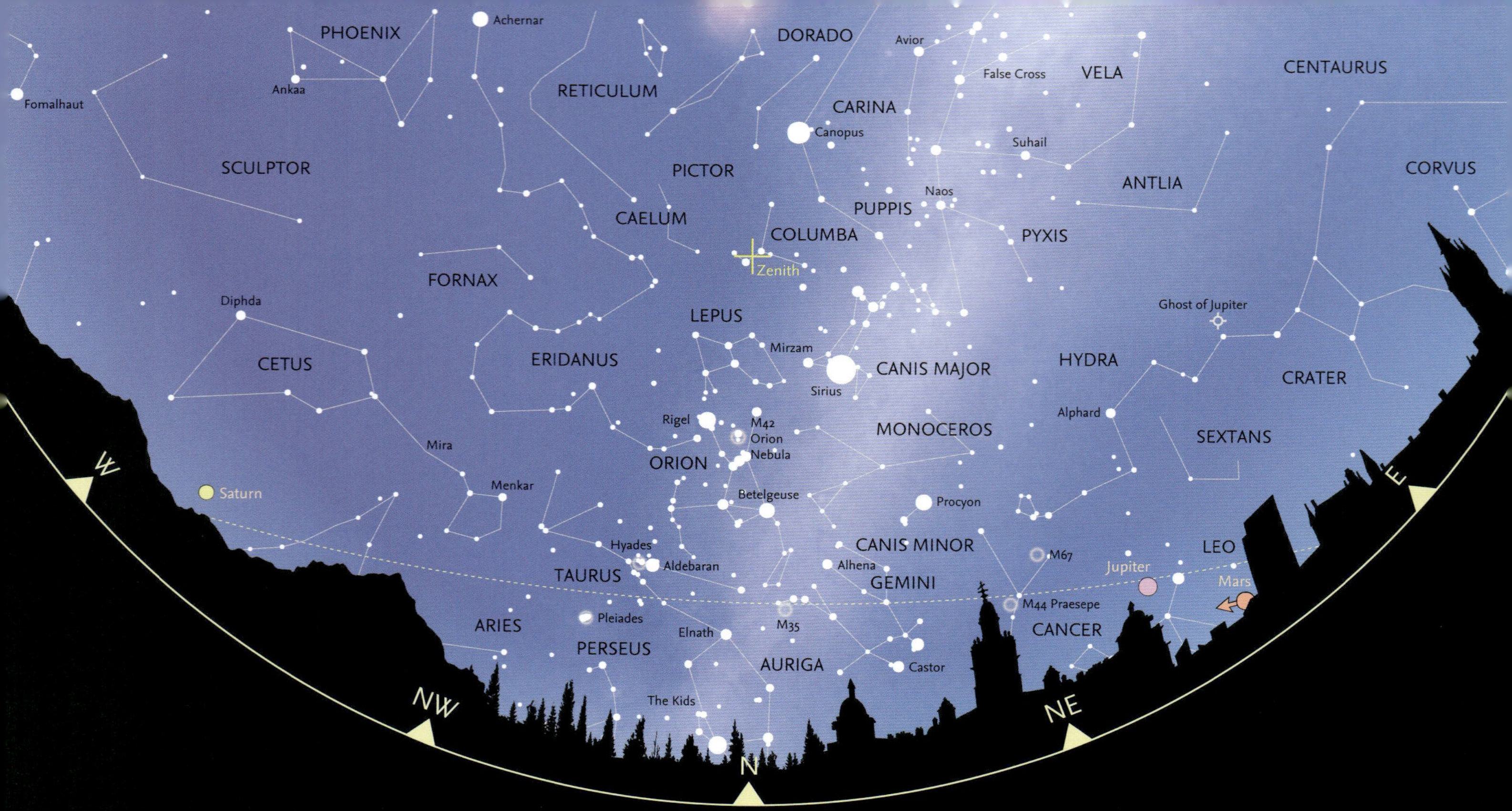

January – Looking North

1 January	23:00	(00:00 DST)
15 January	22:00	(23:00 DST)
1 February	21:00	(22:00 DST)

January – Looking North

The northern sky is dominated by ***Orion***, prominent during the summer months and visible at some time during the night. It is highly distinctive, with a line of three stars that form Orion's Belt. To most observers, the bright star ***Betelgeuse*** (α Orionis) shows a reddish tinge; it is a red supergiant star, in contrast to the brilliant bluish-white ***Rigel*** (β Orionis). The three stars of the belt lie directly south of the celestial equator. A vertical line of three 'stars' forms the 'sword' that hangs south of the 'belt'. With good viewing, the central 'star' appears as a hazy spot, even to the naked eye, and is actually the ***Orion Nebula***, a star formation region. Binoculars reveal the four stars of the ***Trapezium*** cluster, which illuminate the nebula.

The Belt points down to the northwest towards ***Taurus*** (the Bull) and orange-tinted ***Aldebaran*** (α Tauri). Close to Aldebaran, there is a conspicuous 'V' of stars, called the ***Hyades*** cluster. Aldebaran lies much closer to the Earth than the Hyades, it just so happens to share the same line of sight. Farther along, the same line from Orion passes above a bright cluster of stars, the ***Pleiades***, or Seven Sisters. Even the smallest pair of binoculars reveals this as a beautiful group of bluish-white stars. Another conspicuous star in Taurus, ***Elnath*** (β Tauri), lies directly above Orion and forms a big 'V' with Aldebaran.

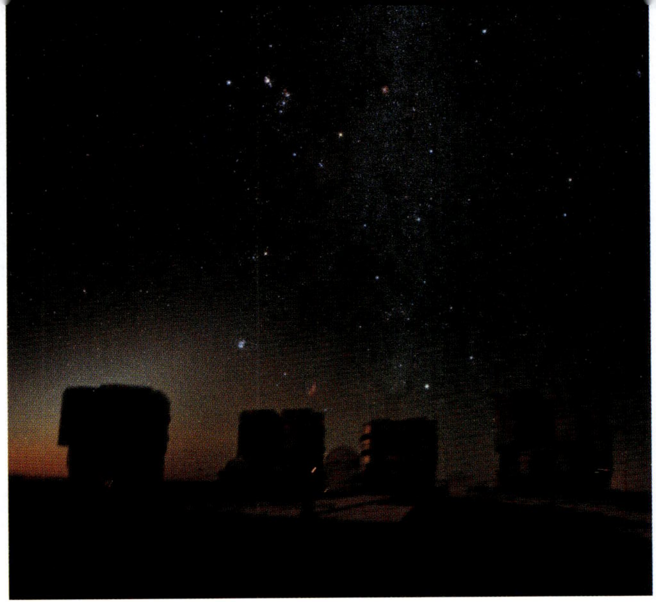

Orion the Greek hunter, over the Very Large Telescope in the Atacama Desert, Chile.

The Moon's phases for January 2027

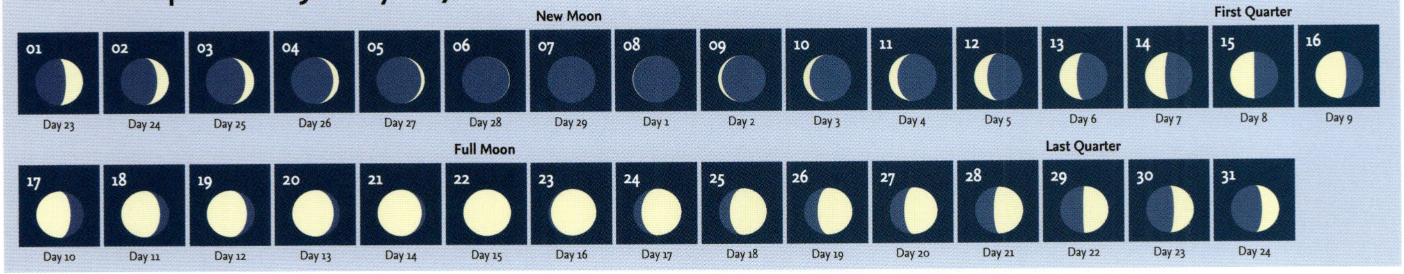

January – Moon and Planets

The Moon

The year starts with the waning crescent Moon 2.8°S of the star ***Spica*** in the early hours of New Year's Day. Three days later the Moon lies 0.3°S of ***Antares***, visible at dawn. On 7 January the New Moon reaches its farthest point from Earth at a distance of 406,610 km. First Quarter Moon occurs on 15 January and 18 January the young star cluster ***Pleiades*** lies 1.1°S of the waxing gibbous Moon. The Moon is at its closest point to Earth (357,285 km) on 21 January. That night it lies 4.4°S of the star ***Pollux*** and becomes Full the following day. On 24 January, a few hours before sunrise, the waning gibbous Moon sits 1.5°S of ***Jupiter*** (mag. -2.5) and 1.5°S of ***Regulus***. On 28 January the Moon passes close to ***Virgo***'s brightest star ***Spica*** appearing 2.9°S. The day after brings a Last Quarter Moon and the month ends with ***Antares*** sitting 0.4°N of the waning crescent Moon.

The Planets

Mercury starts the month at solar conjunction, it sets with the Sun and then gradually trails the Sun, becoming visible at sunset and dimming from mag. -1.3 to -0.8 at the end of January. ***Venus*** leads the Sun at dawn at the start of the year, reaching eastern elongation on 3 January at mag. -4.5 and dimming to mag. -4.3 at the end of the month. ***Mars*** rises after 21:00 in ***Leo*** at mag. -0.1, it enters retrograde (westward) motion on 10 January and completes the month at mag. -0.9. ***Jupiter*** lies in ***Leo***, rising ahead of Mars and best seen later in the evening. Our largest planet brightens from mag. -2.4 to mag. -2.5, passing 1.5°N of the waning gibbous Moon on 24 January. ***Saturn*** lies in ***Cetus*** during the month. It's visible until an hour before midnight and dims from mag. 0.7 to mag. 0.8. ***Uranus*** sits in ***Taurus*** most of the night, starting at mag. 5.6 and dimming to 5.7. ***Neptune*** is in ***Pisces*** at mag. 7.8.

The path of the Sun and the planets along the ecliptic in January.

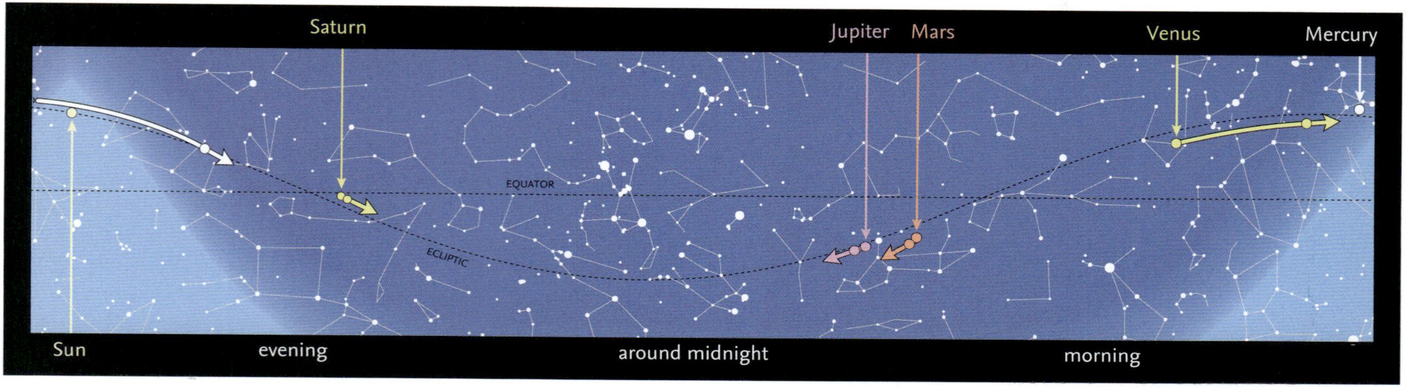

Calendar for January

01	00:29	Spica 2.8°N of Moon
03	02:32	Earth at perihelion (147,099,586 km = 0.9833 AU)
03	18:21	Venus at greatest elongation (47.0°W, mag. -4.5)
04	16:34	Antares 0.3°N of the Moon; an occultation will be visible from parts of New Zealand, Fiji, Samoa, Tonga
07	08:10	Moon at apogee = 406,610 km
07	20:24	New Moon
10	12:58	Mars enters retrograde motion
15	20:34	First Quarter Moon
18	08:46	Pleiades 1.1°S of the Moon
21	21:49	Moon at perigee = 357,285 km
21	23:18	Pollux 4.4°N of the Moon
22	12:17	Full Moon
24	00:24	Jupiter (mag. -2.5) 1.5°N of the Moon
24	09:19	Regulus 1.5°N of the Moon
28	07:55	Spica 2.9°N of the Moon
29	10:55	Last Quarter Moon
31	22:48	Antares 0.4°N of the Moon

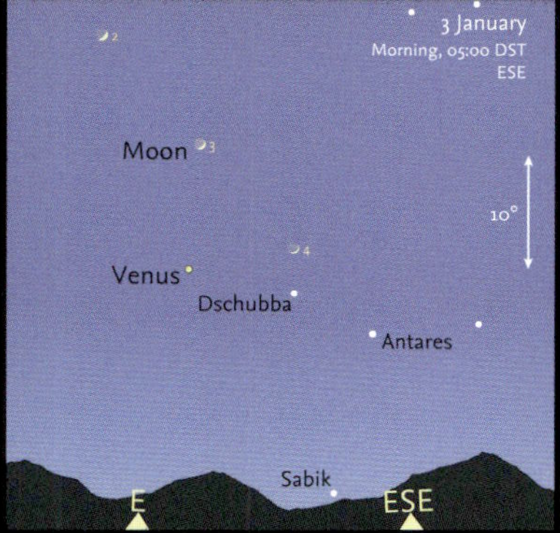

3 January • *The waning crescent Moon passes by Venus.*

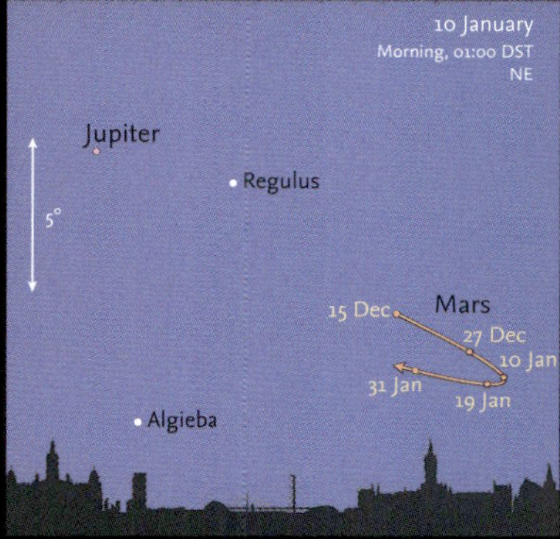

10 January • *Mars enters retrograde (westward) motion, Jupiter and Regulus are close by.*

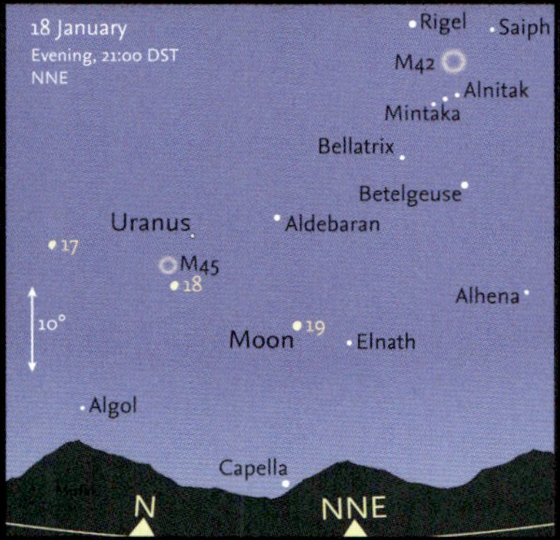

18 January • *The waxing gibbous Moon close to the Pleiades (M45) and Uranus.*

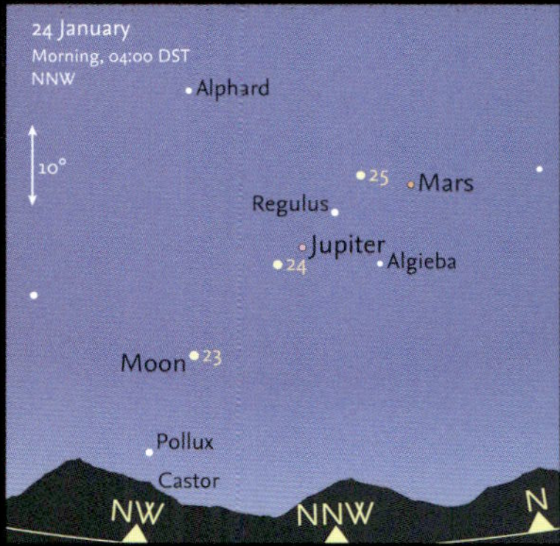

24 January • *Jupiter and the waning gibbous Moon together in the sky, Mars further east.*

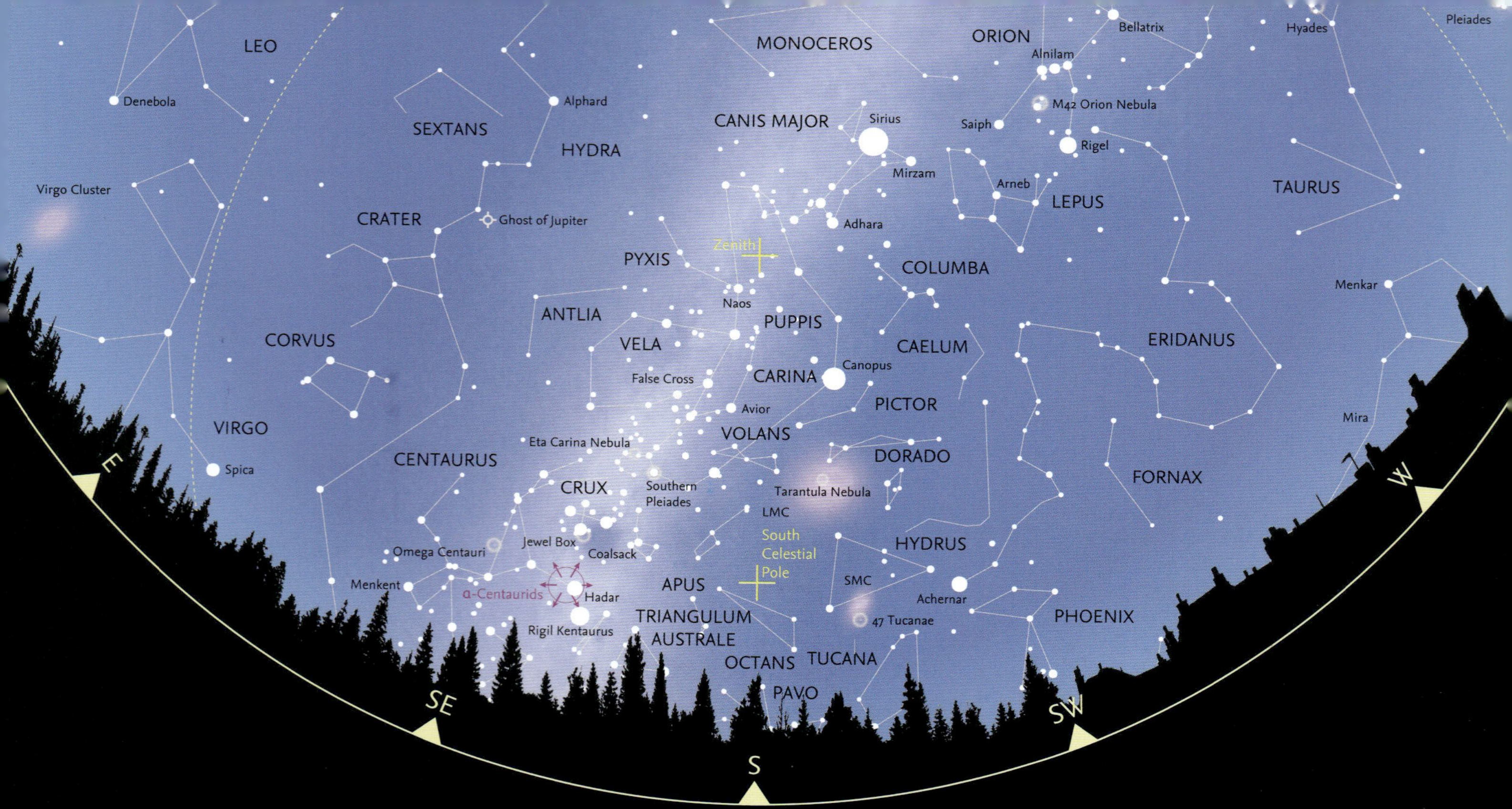

February – Looking South

1 February	23:00	(00:00 DST)
15 February	22:00	(23:00 DST)
1 March	21:00	(22:00 DST)

February – Looking South

The whole of ***Centaurus*** is now clear of the horizon in the southeast, where, below it, the constellation of ***Lupus*** is beginning to be visible. ***Triangulum Australe*** is now higher in the sky and easier to see. Both ***Crux*** and the ***False Cross*** are also higher and more visible. The ***Large Magellanic Cloud*** (LMC) and brilliant ***Canopus*** are now almost due south, with the constellation of ***Puppis*** at the zenith. Both ***Hydrus*** and the ***Small Magellanic Cloud*** (SMC) are lower in the sky, as is ***Phoenix***, which is much closer to the horizon in the west. Next to it, the constellation of ***Tucana*** is also lower although the globular cluster ***47 Tucanae*** remains readily visible as does ***Achernar*** (α Eridani).

Meteors

The ***Centaurid*** shower (which actually consists of two separate streams: the α- and β-Centaurids), whose radiants lie near α and β Centauri (Rigil Kentaurus and Hadar), respectively – continues in February. The shower reaches a low maximum of around six meteors per hour on 8 February, when the Moon is a waxing crescent setting in the evening, so observations are best after 20:00. Another weak shower, the ***γ-Normids***, begins to be active on 25 February, but the meteors are difficult to differentiate from sporadics. It peaks on 14 March, with a similar hourly rate to the Centaurids.

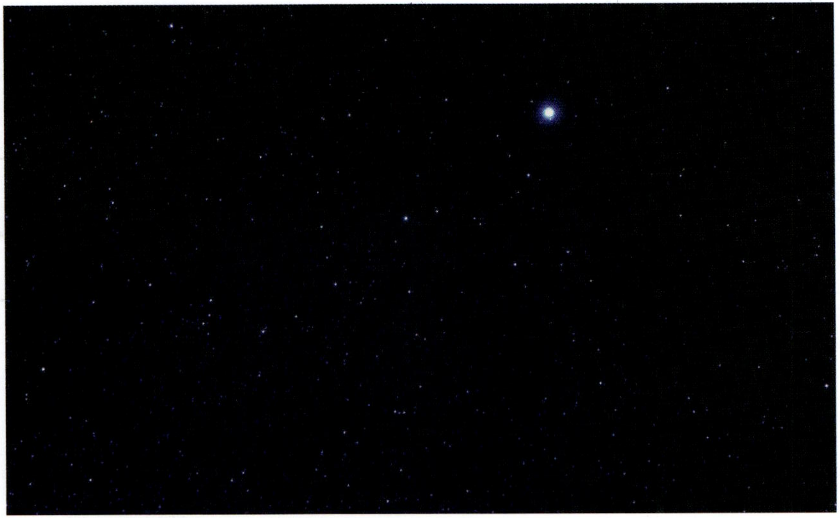

The constellation Carina and its brightest star Canopus.

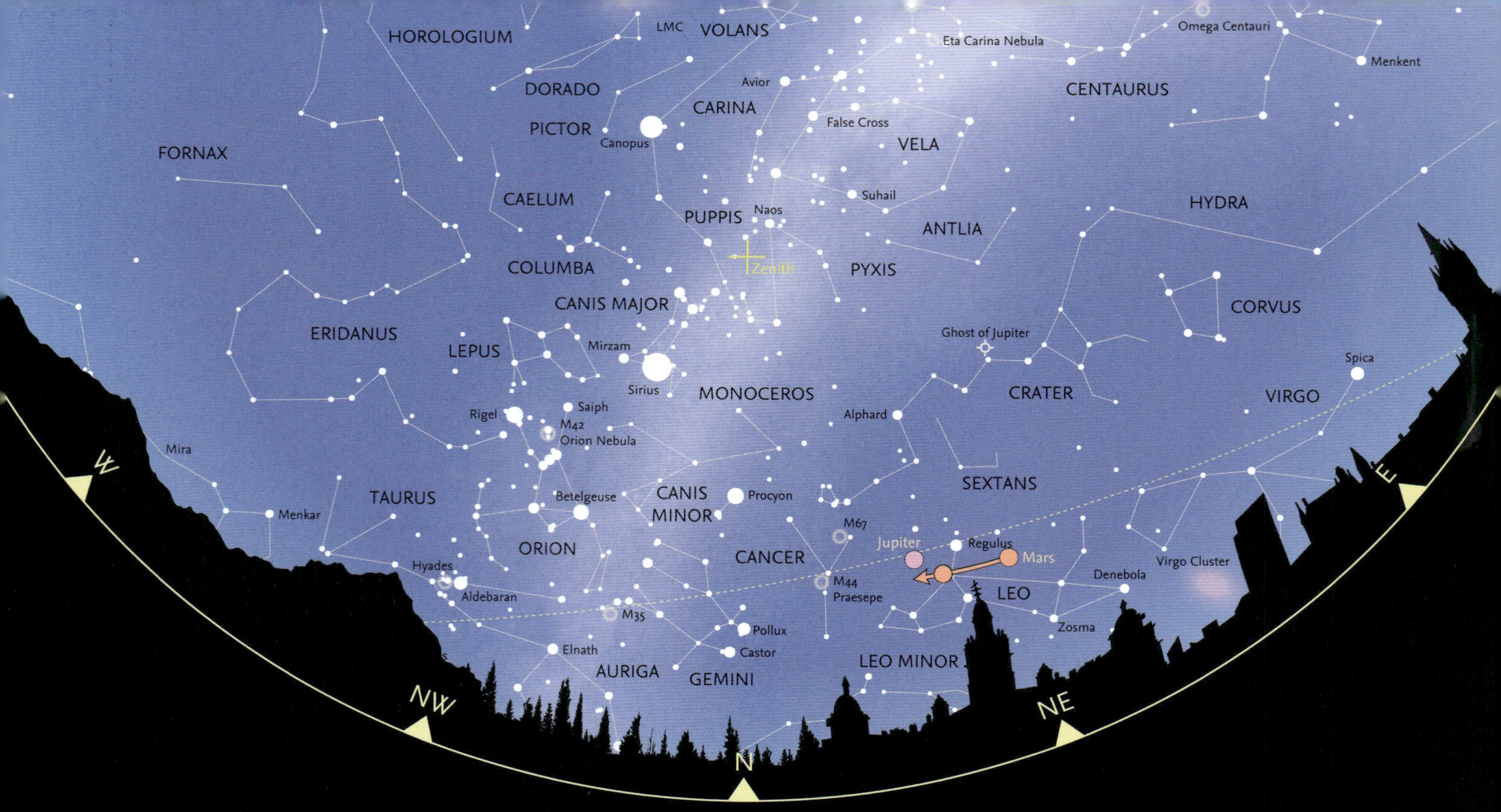

February – Looking North

1 February	23:00	(00:00 DST)
15 February	22:00	(23:00 DST)
1 March	21:00	(22:00 DST)

February – Looking North

The constellation of ***Gemini***, with the pair of stars ***Castor*** and ***Pollux*** (α and β Geminorum) is now due north. Pollux is higher in the sky (farther towards the zenith), and above it is ***Procyon*** (α Canis Minoris), halfway to the zenith. Pollux (β Geminorum) is the brighter of the two (mag. 1.2), it lies in the ecliptic and is occasionally occulted by the Moon. Castor (α Geminorum), the fainter star (mag. 1.9), is remarkable because it is actually a multiple system, consisting of no fewer than six individual stars arranged in three binary (paired) systems.

Still higher is ***Sirius***, the brightest star in the sky (at mag. -1.4) and in the constellation of ***Canis Major***. The faint constellation of ***Cancer***, with its most noticeable feature, the cluster ***Praesepe***, lies to the east of Gemini. Above it, and directly east of Procyon, is the distinctive asterism forming the head of ***Hydra***, the largest of all 88 constellations, the whole of which is now visible stretching across the sky towards the east. The constellation of ***Taurus***, with orange ***Aldebaran*** (α Tauri) is still clearly seen in the west. By contrast, ***Auriga*** is much lower towards the horizon and brilliant ***Capella*** (α Aurigae) is extremely low and visible only early in the night. The faintest stretch of the Milky Way runs from Auriga in the northwest up towards the zenith, passing through Gemini and the constellation of ***Monoceros***. In the northeast, the zodiacal constellation of ***Leo*** and bright ***Regulus*** (α Leonis) are clearly seen.

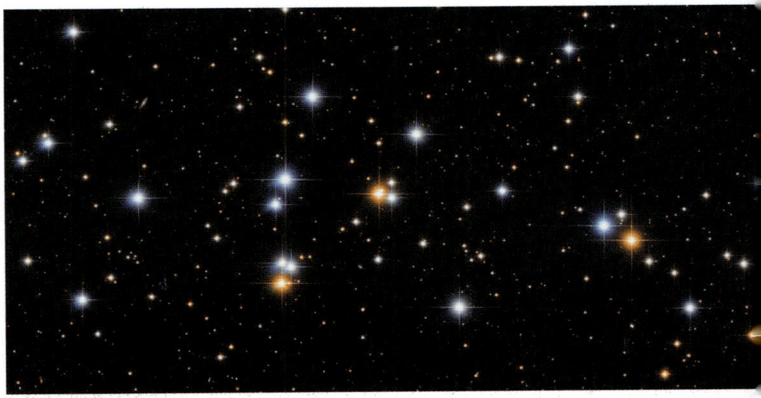

The Beehive Cluster, an open cluster of stars.

The Moon's phases for February 2027

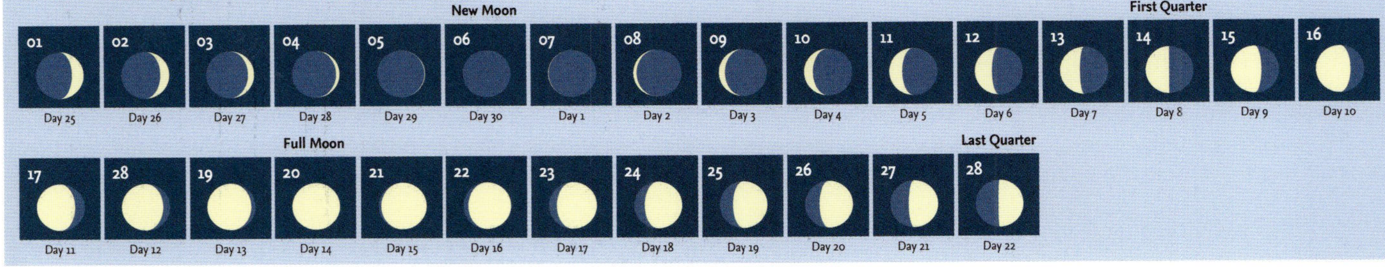

February – Moon and Planets

The Moon

The lunar phase cycle resets on 6 February with a New Moon. On 8 February ***Mercury*** (mag. 0.2) lies 0.4°N of the slim waxing crescent Moon. The First Quarter Moon arrives 14 February and the ***Pleiades*** can be seen 1.2°S of the Moon. The waxing gibbous Moon passes 4.4°S of ***Pollux*** on 18 February and two days later ***Jupiter*** (mag. -2.5) sits 1.2°N of the Full Moon, with ***Regulus*** 1.5°N. The waning gibbous Moon passes 2.9°S of ***Spica*** on 24 February and reaches Last Quarter on 28 February. On the same day the red supergiant ***Antares*** can be seen 0.4°N of the Moon, appearing low in the southeast a few hours before sunrise.

The Planets

Mercury is visible just after sunset at the beginning of the month at mag. -0.7, reaching eastern elongation on 3 February and switching to the early morning after 18 February, dimming to mag. 5.3 and eventually brightening to mag. 1.4 towards the end of February. It passes 0.4°N of the waxing crescent Moon on 8 February (at mag. 0.2). ***Venus*** continues to lead the Sun at dawn at mag. -4.3 and dims to mag. -4.1 at the end of the month. ***Mars*** rises several hours after sunset in ***Leo*** at mag. -0.9, it reaches opposition on 19 February at mag. -1.3, sits 3.8°N of ***Regulus*** on 22 February and ends the month at mag. -1.2. ***Jupiter*** lies in Leo with Mars and reaches opposition on 11 February at mag. -2.6. On 20 February Jupiter (mag. -2.5) lies 1.2°N of the Full Moon. ***Saturn*** moves through ***Cetus*** at the beginning of the month and arrives at ***Pisces*** from 23 February, visible in the early evening at mag. 0.8. ***Uranus*** sits in ***Taurus*** most of the night, at mag. 5.7. It ends retrograde motion on 8 February. ***Neptune*** is in Pisces at mag. 7.8 close to Saturn.

The path of the Sun and the planets along the ecliptic in February.

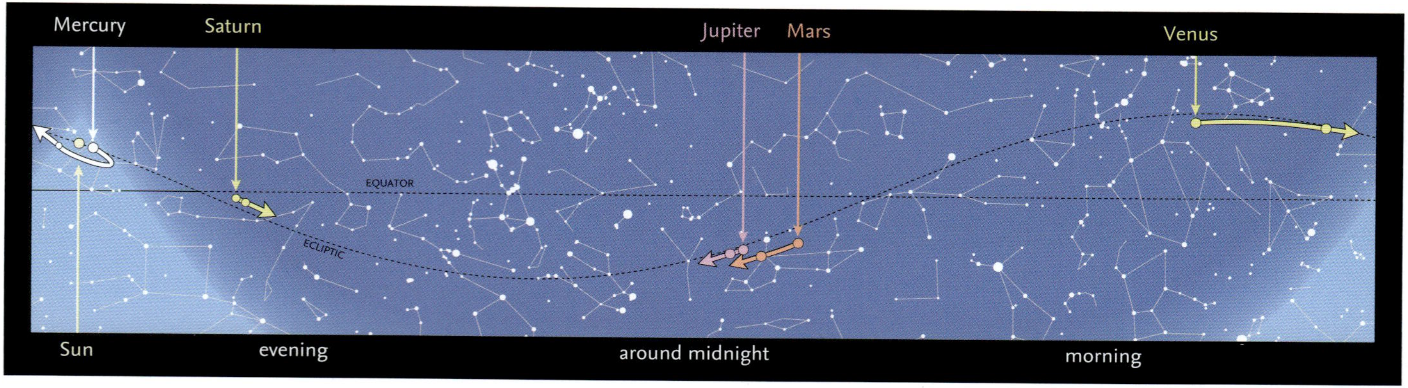

Calendar for February

03	05:11	Mercury at greatest elongation (18.3°E, mag. -0.5)
03	13:31	Moon at apogee = 406,189 km
06	15:56	New Moon
06	15:59	Annular solar eclipse (southern Chile, Argentina, Atlantic); partial eclipse (South America, Antarctica, west and south Africa)
08		Alpha (α) Centaurid meteor shower maximum
08	03:36	Mercury (mag. 0.2) 0.4°N of the Moon
08	11:47	Uranus ends retrograde motion
11	00:21	Jupiter (mag. -2.6) at opposition
14	07:58	First Quarter Moon
14	16:13	Pleiades 1.2°S of the Moon
18	09:50	Pollux 4.4°N of the Moon
19	07:30	Moon at perigee = 361,015 km
19	15:44	Mars (mag. -1.3) at opposition
20	05:46	Jupiter (mag. -2.5) 1.2°N of the Moon
20	20:33	Regulus 1.5°N of the Moon
20	23:12	Penumbral lunar eclipse; visible from the Americas, Africa, Asia, Russia, Antarctica and Europe
20	23:23	Full Moon
22	04:31	Mars (mag. -1.3) 3.8°N of Regulus
24	17:14	Spica 2.9°N of the Moon
28	05:16	Last Quarter Moon
28	06:09	Antares 0.4°N of the Moon; an occultation will be visible from Argentina, Brazil, Bolivia, Chile, Peru, Paraguay, Uruguay, Falkland Islands

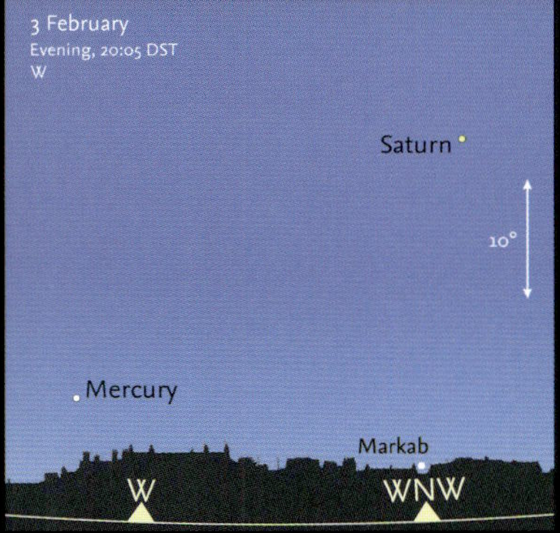

3 February • *Mercury at greatest eastern elongation.*

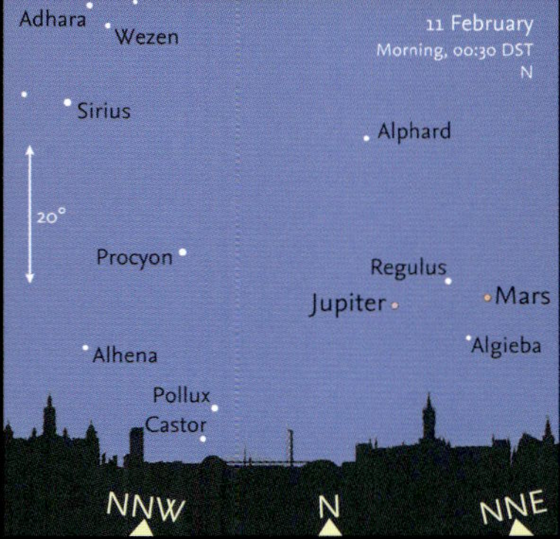

11 February • *Jupiter at opposition due south, with Mars close by.*

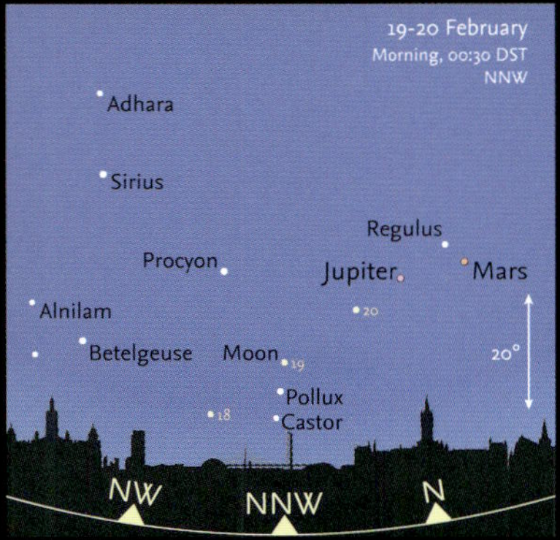

19–20 February • *Mars at opposition, the following day Jupiter sits next to the Full Moon.*

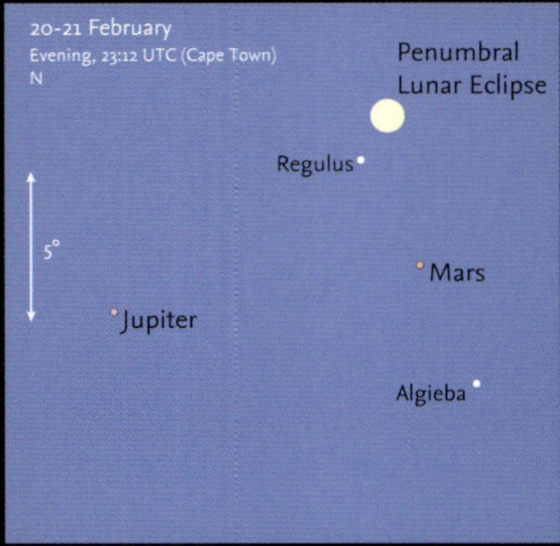

20–21 February • *Penumbral lunar eclipse, Mars and Jupiter nearby.*

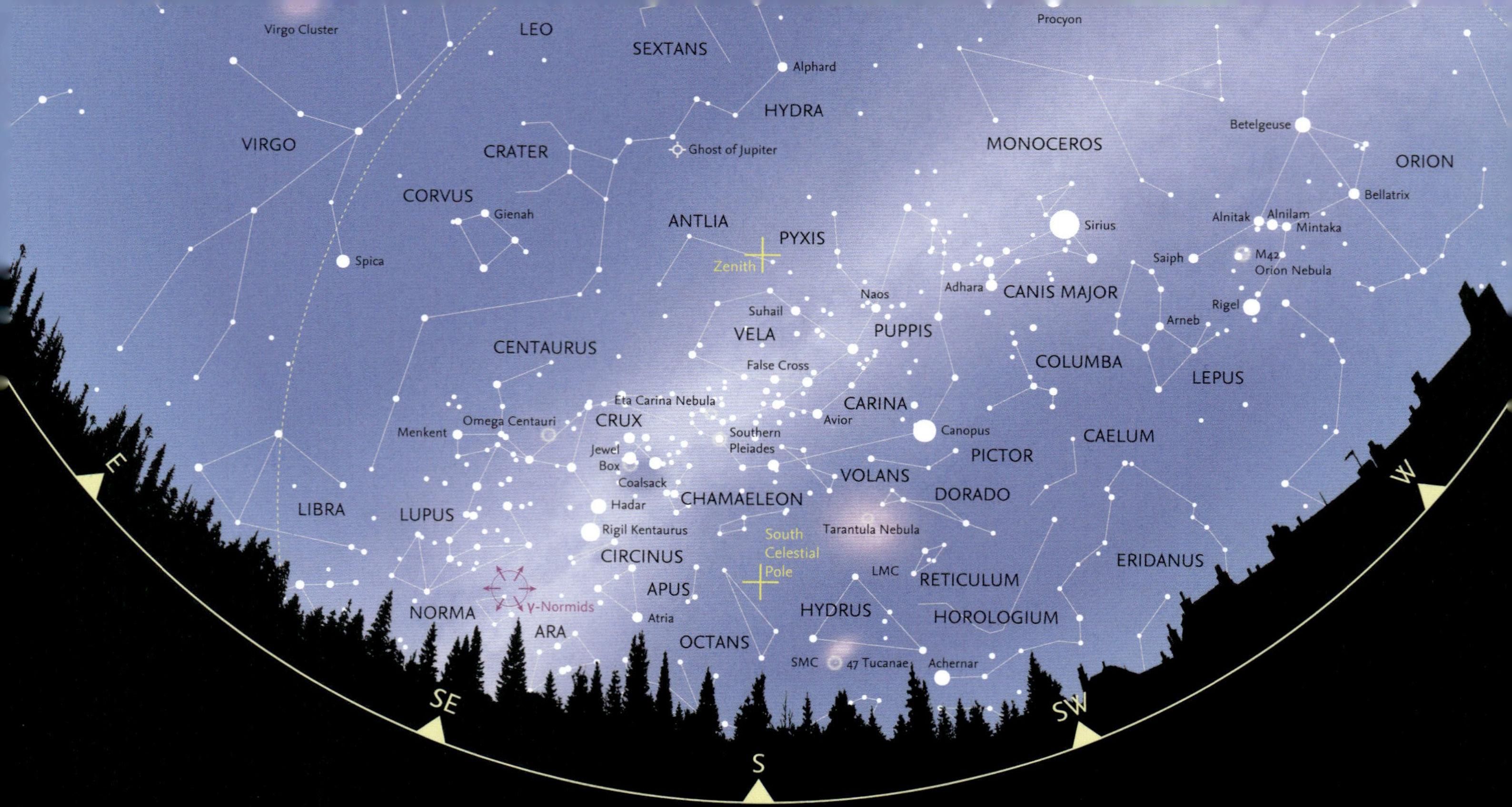

March – Looking South

1 March	23:00	(00:00 DST)
15 March	22:00	(23:00 DST)
1 April	21:00	(22:00 DST)

March – Looking South

The Sun crosses the celestial equator on 20 March, at the autumnal equinox, when day and night are of almost equal length, and the southern season of autumn is considered to begin. The hours of daylight change most rapidly around the equinoxes in March and September.

Scorpius is beginning to become visible in the eastern sky, including the Cat's Eyes, the pair of stars ***Shaula*** and ***Lesath*** (λ and υ Scorpii, respectively) at the end of the 'sting'. Above Scorpius, the whole of the constellation of ***Lupus*** is now easy to see. The magnificent globular cluster of ***Omega Centauri*** is now readily visible, northeast of ***Crux***. The ***Coalsack Nebula*** (a dark nebula of interstellar dust obscuring starlight) and the denser region of the Milky Way in ***Carina***, together with the ***Eta Carinae Nebula*** (containing massive stars considerably hotter than the Sun) and the ***Southern Pleiades*** are well placed for observation. The ***False Cross*** on the ***Carina/Vela*** border is now high in the sky, between the South Celestial Pole and the zenith. Brilliant ***Canopus*** (α Carinae) is only slightly lower towards the west, above the ***Large Magellanic Cloud*** (LMC) and the striking ***Tarantula Nebula***, a star-forming region visible to the naked eye. ***Achernar*** (α Eridani), the ***Small Magellanic Cloud*** (SMC) and ***47 Tucanae*** are considerably lower, but still clear of the horizon. ***Peacock*** (α Pavonis), a binary star system in the constellation ***Pavo***, skims the horizon, just east of south. ***Orion*** is now visibly getting lower in the west and is being followed by ***Sirius*** and ***Canis Major***.

Meteors

The meteor shower in March is the ***γ-Normids***, which have a low rate, and are thus difficult to differentiate from sporadics. However, they exhibit a very sharp peak a day or so on either side of maximum on 14 March, when the Moon is almost First Quarter and interfering with observations. The faint constellation of ***Norma*** rises early in the night, but most meteors are likely to be seen (away from the radiant) in the hours after midnight, after the Moon sets.

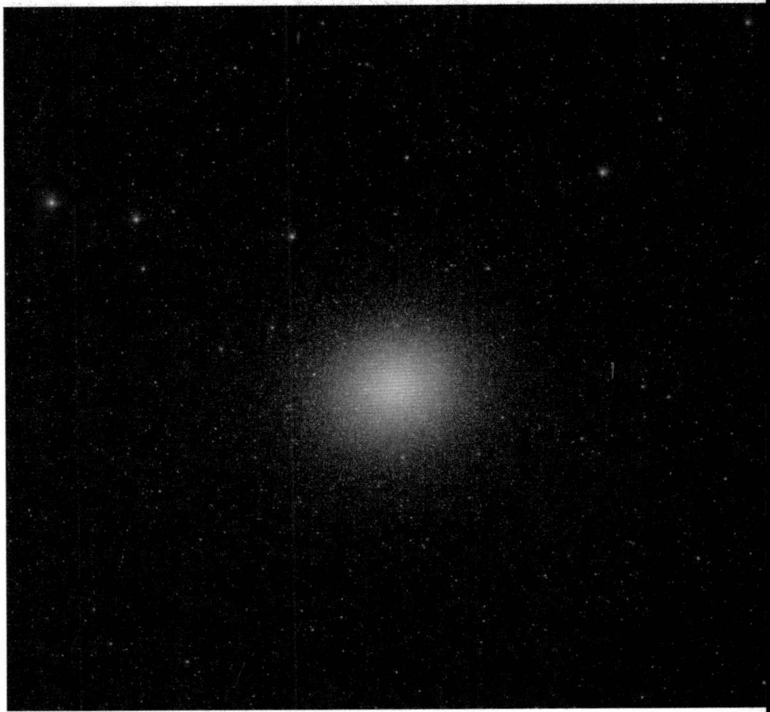

The spectacular globular cluster, Omega Centauri.

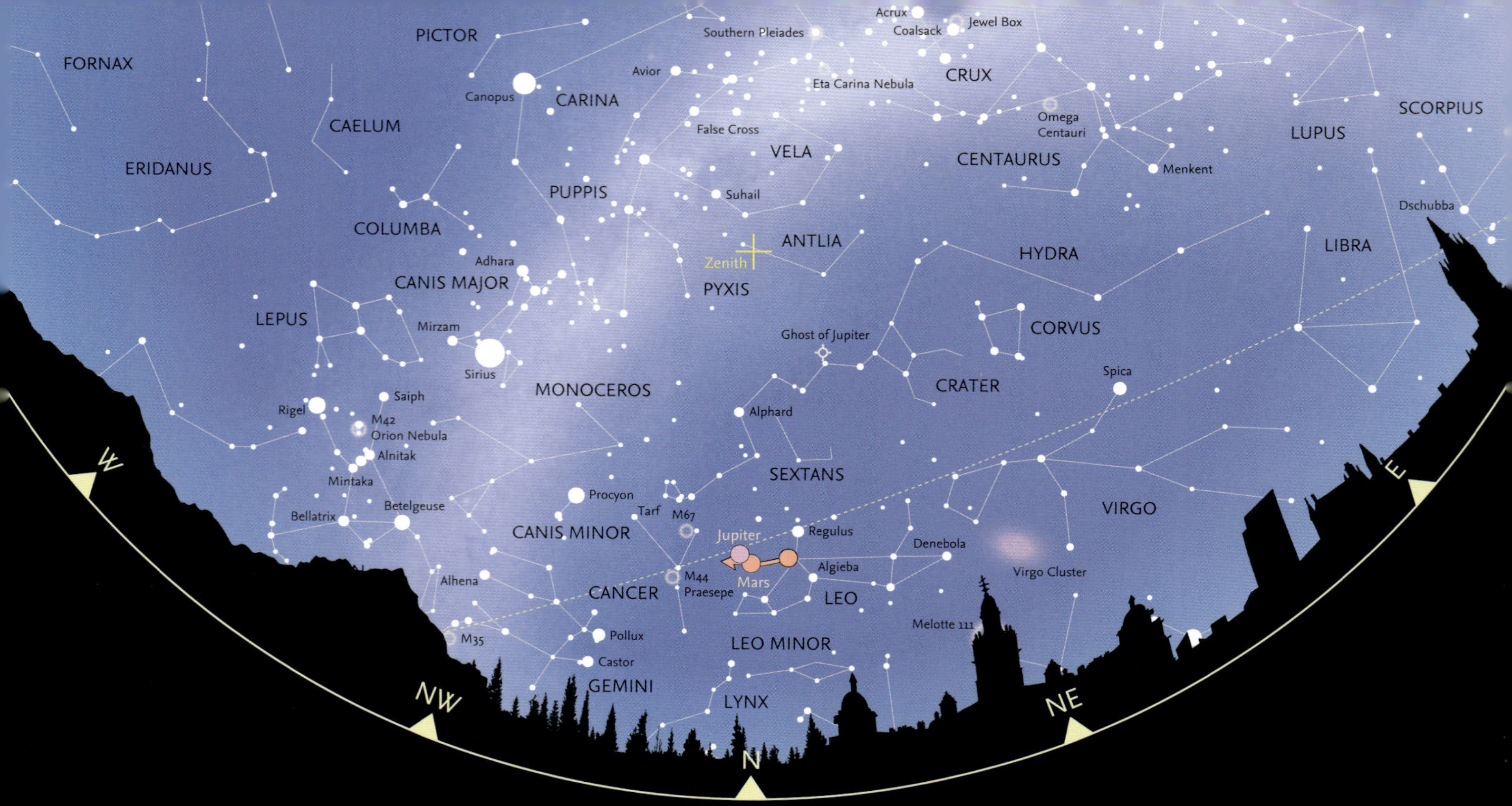

March – Looking North

1 March	23:00	(00:00 DST)
15 March	22:00	(23:00 DST)
1 April	21:00	(22:00 DST)

March – Looking North

Almost due north is the constellation of ***Leo***, with the 'upside-down backwards question mark' (or the Sickle) of bright stars forming the head of the mythological lion. ***Regulus*** (α Leonis) – the 'dot' of the question mark or the 'handle' of the sickle and the brightest star in Leo – lies very close to the ecliptic and is one of the few first-magnitude stars that may be occulted by the Moon. ***Denebola*** (β Leonis), a variable star, marks the tail of the lion. The Moon often passes between Regulus and ***Algieba*** (γ Leonis). To the west lies the faint zodiacal constellation of ***Cancer*** (the Crab), its brightest star is β Cancri (also called ***Tarf***), an orange giant around 50 times the size of the Sun. Near the centre of the constellation lies an open cluster, M44 or ***Praesepe*** (the Manger but also known as the Beehive). On a clear night this cluster of around 1,000 stars, known since antiquity, is just a hazy spot to the naked eye, but appears in binoculars as a group of dozens of individual stars. The constellations of ***Gemini*** and ***Orion*** are now getting low in the west, but above them, both ***Procyon*** in ***Canis Minor*** and the constellation of ***Canis Major*** remain clear to see. The whole of ***Hydra*** (the largest constellation) now sprawls right across the northern and eastern skies, and the three small constellations of ***Sextans***, ***Crater*** and ***Corvus*** are readily visible. The small constellation of ***Antlia*** is at the zenith. In the east, ***Arcturus*** (α Boötis) – the brightest star in the northern hemisphere of the sky – is becoming visible, and climbs higher during the night.

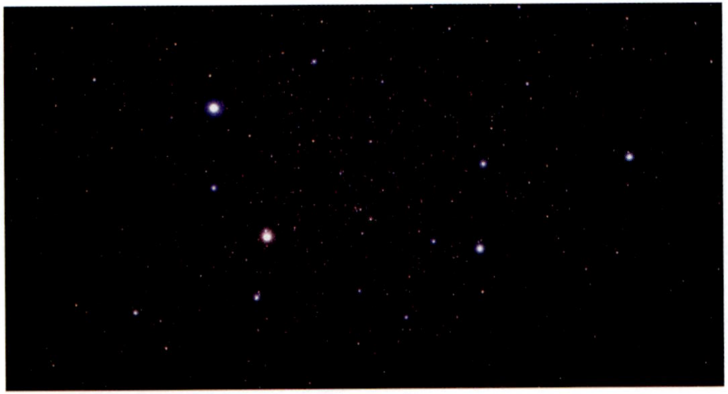

Leo the lion, with its 'backwards question mark'.

The Moon's phases for March 2027

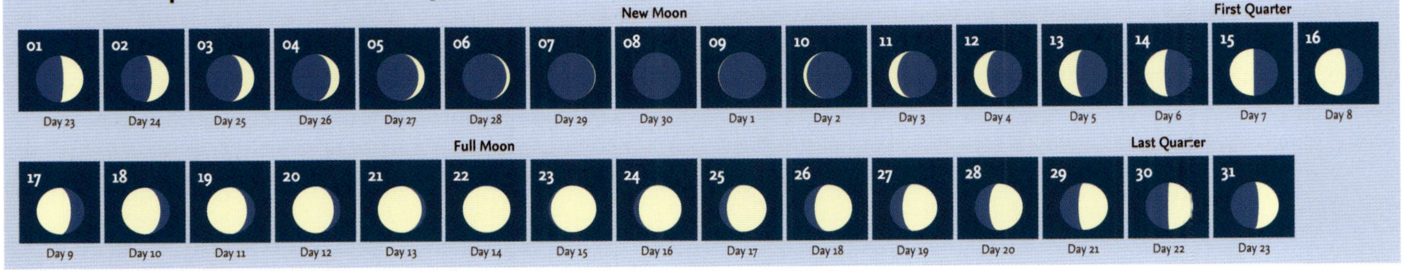

March – Moon and Planets

The Moon

On 4 March ***Venus*** (mag. -4.1) can be found 2.1°N of the waning crescent Moon – best seen an hour before sunrise the following day. On 6 March the thin crescent Moon passes 1.2°S of ***Mercury*** (mag. 0.6), also at dawn. The New Moon is on 8 March and five days later the ***Pleiades*** star cluster is placed 1.1°S of the Moon. The First Quarter Moon on 15 March is followed two days later by a close approach with ***Pollux***, sitting 4.5°N of the waxing gibbous Moon. On 19 March ***Jupiter*** (mag. -2.4) lies 0.9°N of the bright waxing gibbous Moon, with ***Mars*** (mag. -0.7) 4.3°N of the Moon. The following day the Moon appears 1.5°S of ***Regulus*** and it reaches a Full Moon on 22 March. ***Spica*** is placed 2.8°N of the Moon on 24 March, ***Antares*** sits closer at 0.3°N of the waning gibbous Moon on 27 March, above the horizon after midnight. The Moon reaches Last Quarter phase on 30 March.

The Planets

Mercury is present in the early morning sky starting the month at mag. 1.2. It climbs to western elongation on 17 March and brightens to mag. -0.1 at the end of the month. ***Venus*** continues to lead the Sun at dawn at mag. -4.1, becoming gradually fainter during the month at mag. -3.9. ***Mars*** is visible after sunset in ***Leo*** at mag. -1.2, dimming to mag. -0.4 at the end of March. ***Jupiter*** lies in Leo with Mars and enters ***Cancer*** from 17 March, visible after sunset, dimming from mag. -2.5 to -2.3. Jupiter (mag. -2.4) lies 0.9°N of the waxing gibbous Moon on 19 March and on the last day of March Mars (mag. -0.4) appears 4.3°N of Jupiter (mag. -2.3). ***Saturn*** stays in ***Pisces***, visible in the early evening at mag. 0.8 and brightening to 0.7. ***Uranus*** is visible in ***Taurus*** until midnight, at mag. 5.7 to 5.8. ***Neptune*** is in Pisces in the early evening at mag. 7.8.

The path of the Sun and the planets along the ecliptic in March.

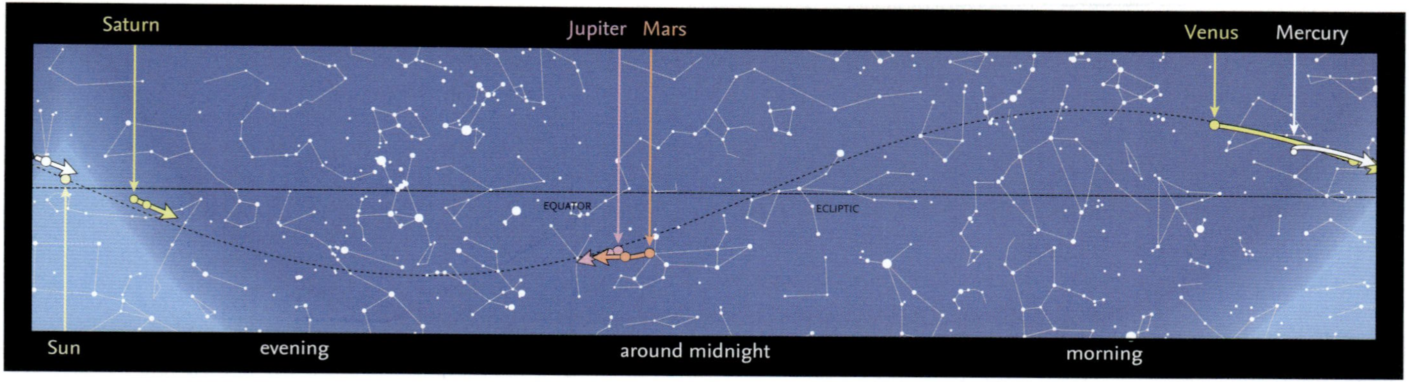

Calendar for March

03	05:41	Moon at apogee = 405,221 km
04	18:59	Venus (mag. -4.1) 2.1°N of the Moon
06	05:28	Mercury (mag. 0.6) 1.2°N of the Moon
08	09:29	New Moon
13	21:43	Pleiades 1.1°S of the Moon
14		Gamma (γ) Normid meteor shower maximum
15	16:25	First Quarter Moon
17	04:20	Mercury at greatest elongation (27.6°W, mag. 0.2)
17	17:40	Pollux 4.5°N of the Moon
19	04:31	Moon at perigee = 366,440 km
19	10:05	Jupiter (mag. -2.4) 0.9°N of the Moon; an occultation will be visible from Australia, New Zealand, Tasmania and southern Papua New Guinea
19	18:33	Mars (mag. -0.7) 4.3°N of the Moon
20	06:13	Regulus 1.5°N of the Moon
20	20:25	Vernal Equinox
22	10:44	Full Moon
24	03:12	Spica 2.8°N of the Moon
27	14:36	Antares 0.3°N of the Moon; an occultation will be visible from Australia, New Zealand, Indonesia, Tasmania and Papua New Guinea
30	00:54	Last Quarter Moon
31	01:33	Moon at apogee = 404,331 km
31	06:00	Mars (mag. -0.4) 4.3°N of Jupiter (mag. -2.3)

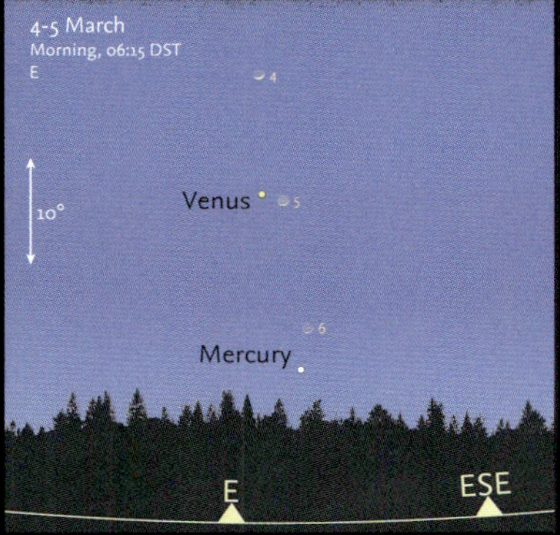

4–5 March • *Venus close to the waning crescent Moon, Mercury low in the sky.*

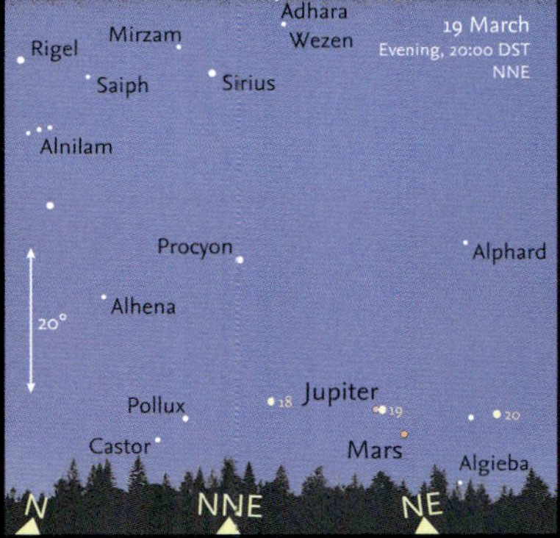

19 March • *Waxing gibbous Moon near Mars and Jupiter.*

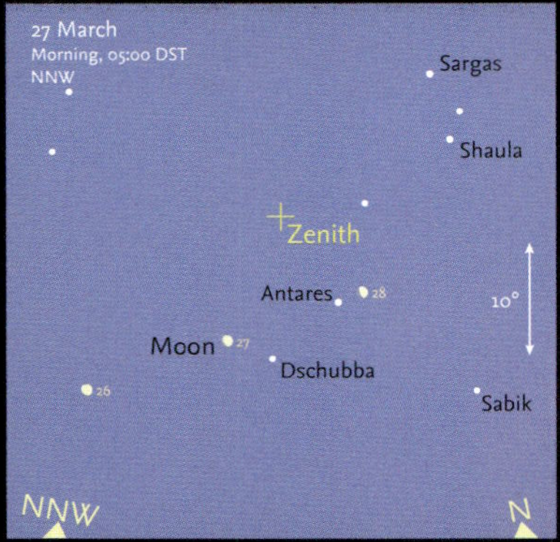

27 March • *Red star Antares close to the waning gibbous Moon.*

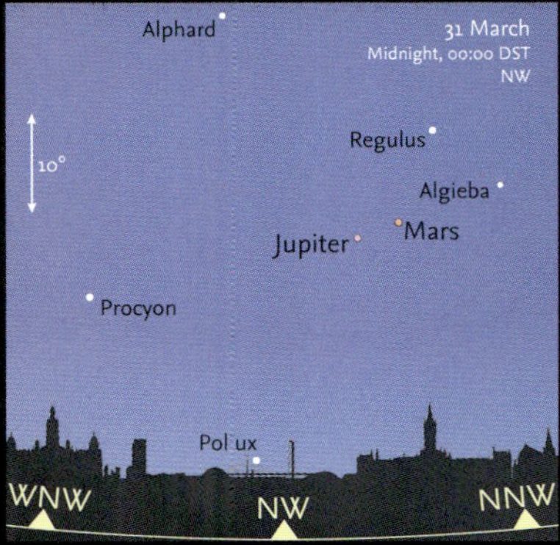

31 March • *Mars and Jupiter with Regulus.*

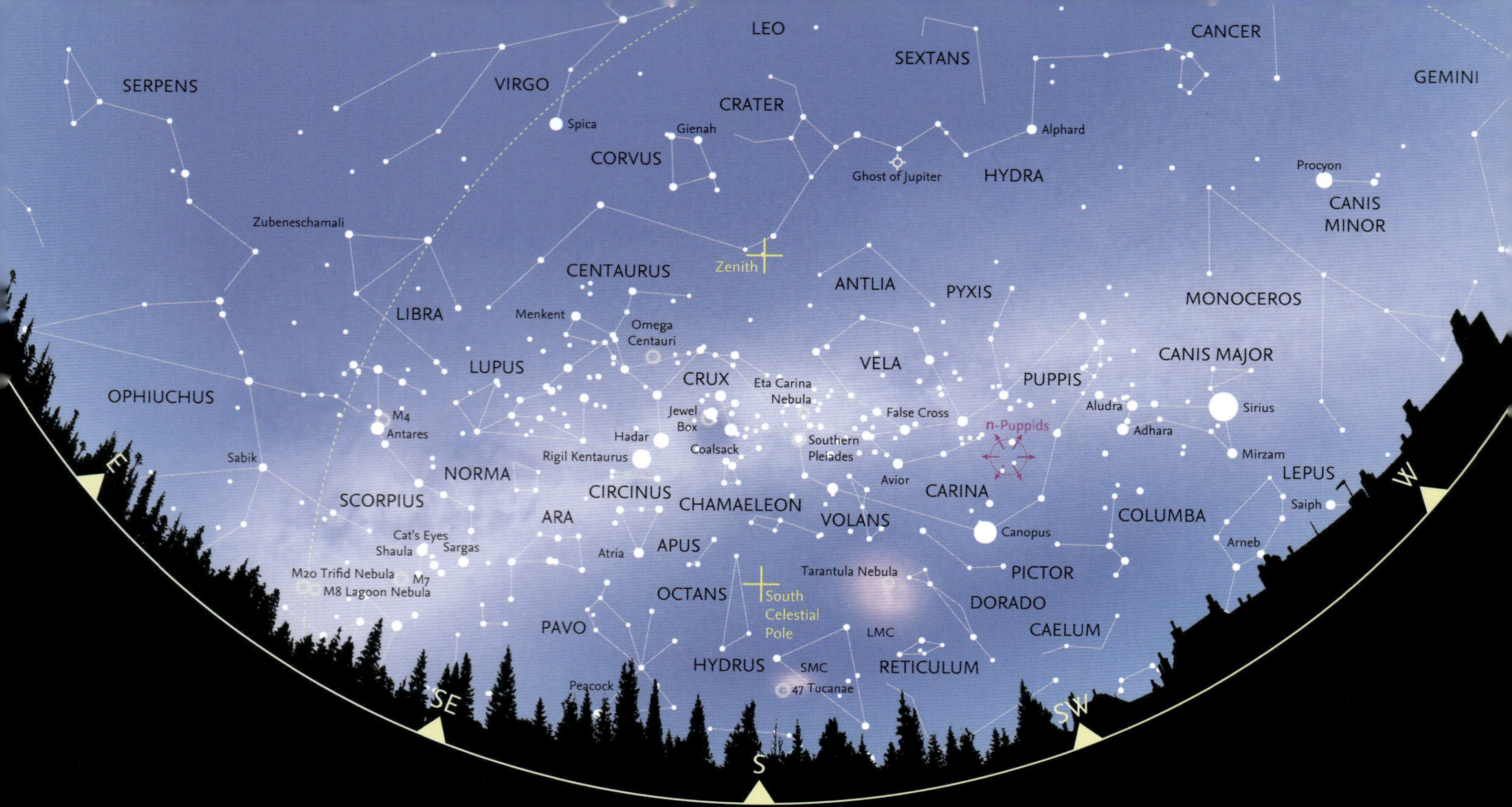

April – Looking South

1 April	23:00	(00:00 DST)
15 April	22:00	
1 May	21:00	

April – Looking South

Daylight Saving Time ends (where used) in Australia and in New Zealand on Sunday 4 April. ***Crux*** is now high in the south, with the two brightest stars of ***Centaurus*** (α and β Centauri, ***Rigil Kentaurus*** and ***Hadar***, respectively) to its east. The magnificent globular cluster, ***Omega Centauri***, is readily visible high in the sky. West of Crux, both the ***Southern Pleiades*** and the ***Eta Carinae Nebula*** are clearly seen, two-thirds of the way towards the zenith. Farther west, the ***False Cross*** is beginning to decline towards the horizon. ***Canopus*** (α Carinae) is even lower, and ***Orion*** has now disappeared below the horizon. ***Canis Major*** and brilliant ***Sirius*** are also descending in the west. ***Peacock*** (α Pavonis) is now slightly higher and more easily visible. Although the ***Large Magellanic Cloud*** (LMC) is roughly as high as the South Celestial Pole, the ***Small Magellanic Cloud*** (SMC) and ***47 Tucanae*** are rather low (but still visible) in the south.

In the east, the whole of ***Scorpius*** is now well clear of the horizon with the neighbouring zodiacal constellation ***Libra*** preceding it along the ecliptic. The dense regions of the Milky Way in ***Sagittarius*** become visible later in the night.

Meteors

Two meteor showers occur in April. The ***π-Puppid*** shower begins on 15 April; however, there is a waning gibbous Moon coinciding with the shower maximum on 24 April so moonlight will hamper observations after midnight. The hourly rate is variable but the meteors tend to be faint. The parent body is the Comet 26P/Grigg-Skjellerup. A second, more prolific, shower, the ***η-Aquariids***, begins on 19 April and continues into May.

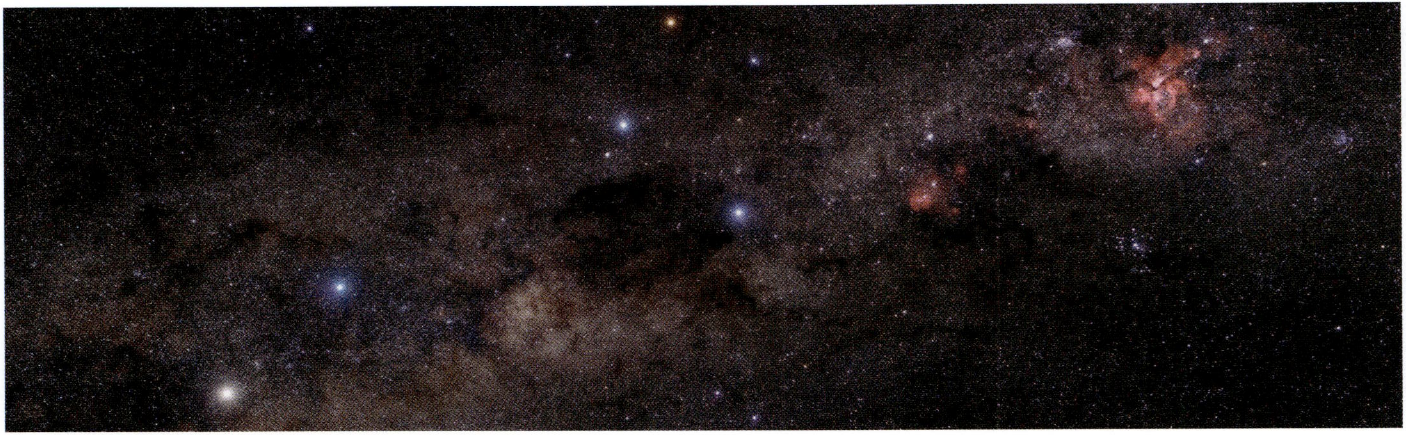

Centaurus, Crux and Carina lying across the band of the Milky Way.

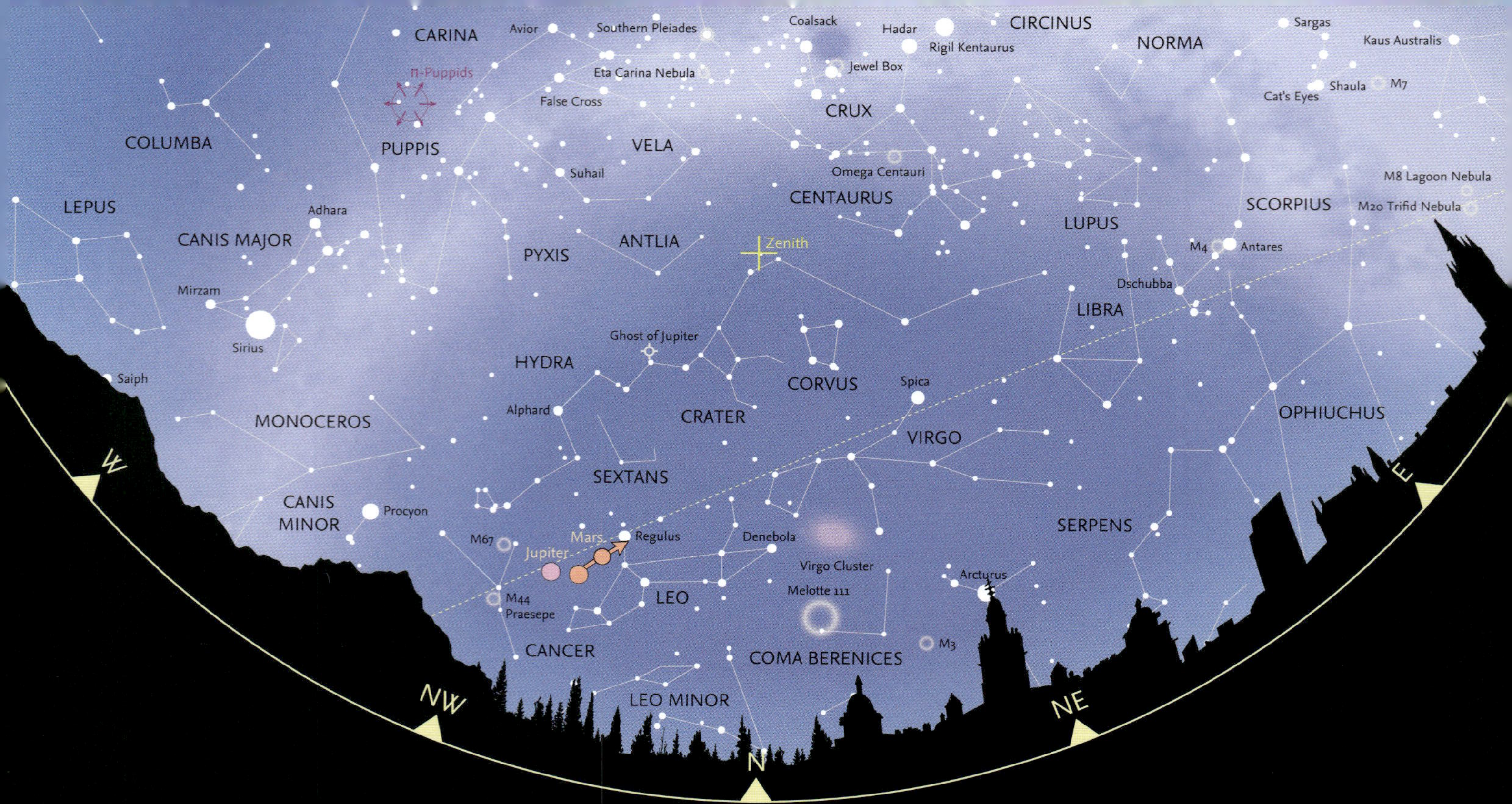

April – Looking North

1 April	23:00	(00:00 DST)
15 April	22:00	
1 May	21:00	

April – Looking North

Leo is the most prominent constellation in the northern sky in April and looks vaguely like the creature after which it is named, albeit upside-down. ***Gemini***, with ***Castor*** and ***Pollux***, and ***Orion*** are low on the horizon in the west, and ***Cancer*** sits next to Leo. To the east of Leo, the whole of ***Virgo***, with ***Spica*** (α Virginis) its brightest star, is clearly visible. The constellation of ***Libra*** is farther east along the ecliptic. Above Leo and Virgo, the complete length of ***Hydra*** is visible, ***Alphard*** (α Hydrae) forms a prominent triangle with ***Regulus*** in Leo and ***Procyon*** in ***Canis Minor***. High in the sky, the small constellations of ***Sextans*** and ***Crater***, together with the rather brighter ***Corvus***, lie between Leo, Virgo and Hydra.

Boötes and ***Arcturus*** are prominent in the northeastern sky, together with the globular cluster ***M3*** in ***Canes Venatici***, visible with binoculars or a small telescope. Between Leo and Boötes lies the constellation of ***Coma Berenices***, the location of the open cluster ***Melotte 111*** (this is sometimes called the Coma Star Cluster and can be confused with the ***Coma Cluster*** of galaxies (Abell 1656). Only about ten of the brightest galaxies in the Coma Cluster are visible with the largest amateur telescopes.

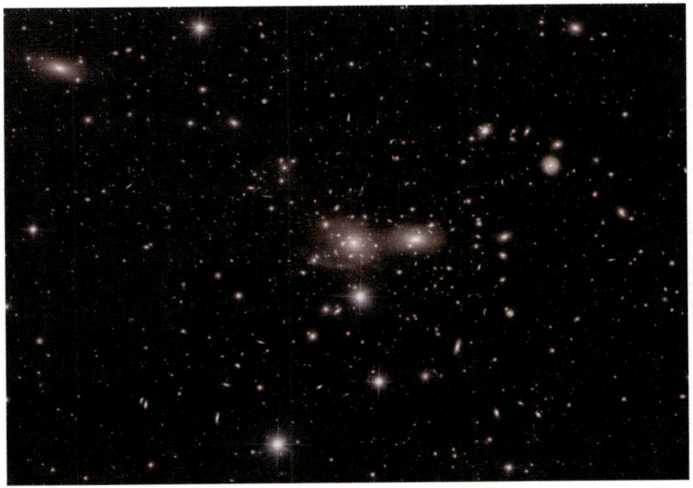
The Coma Cluster, a group of over 1,000 galaxies.

The Moon's phases for April 2027

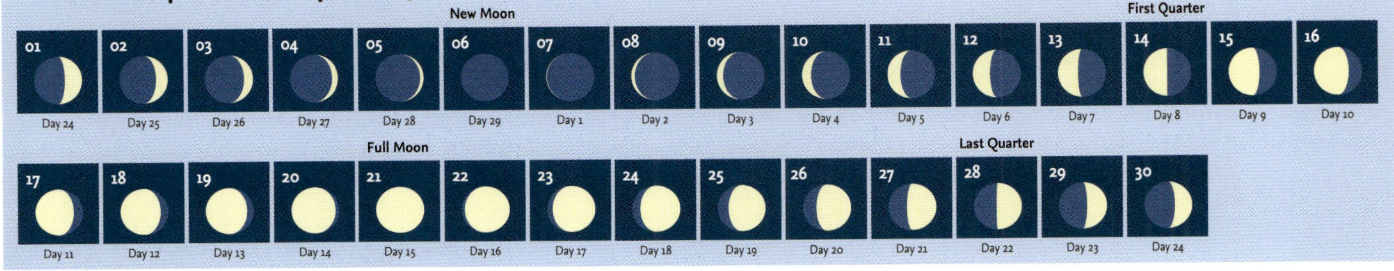

April – Moon and Planets

The Moon

Venus (mag. -3.9) can be seen 3.3°S of the slim waning crescent Moon at dawn on 4 April, two days later the lunar phase cycle starts again with a New Moon. The waxing crescent Moon appears 0.9°N of the ***Pleiades*** on 10 April and on 13 April ***Pollux*** lies 4.7°N of the First Quarter Moon. On 15 April, ***Jupiter*** (mag. -2.2) sits 1.0°N of the waxing gibbous Moon, the day after ***Mars*** (mag. -0.02) is found 3.3°N of the Moon and ***Regulus*** 1.6°N of the Moon. On 20 April the Full Moon passes 2.8°S of ***Spica***, rising later in the evening. ***Antares*** appears close to the waning gibbous Moon, only 0.1°N on 23 April, both above the horizon in the early hours of the following day. The Last Quarter Moon rises on 28 April.

The Planets

Mercury rises with the Sun for most of the month of April. It switches to the evening sky after 28 April, brightening from mag. -0.1 to -2.3. ***Venus*** appears at dawn before the sunrise, easy to see at mag. -3.9. Venus will appear 3.3°S of the waning crescent Moon on 4 April. ***Mars*** ends retrograde motion on 1 April and can be seen in ***Leo*** after sunset at mag. -0.4, dimming to mag. 0.2 as the month progresses. On 16 April Mars (mag. -0.02) lies 3.3°N of the waxing gibbous Moon. ***Jupiter*** lies in ***Cancer***, visible after sunset and dimming from mag. -2.3 to -2.1. It ends retrograde motion on 13 April. On 15 April Jupiter (mag. -2.2) will sit 1.0°N of the waxing gibbous Moon. ***Saturn*** lies in ***Pisces***, setting with the Sun at mag. 0.7. ***Uranus*** is visible in ***Taurus*** in the early evening, at mag. 5.8. ***Neptune*** is in Pisces rising just before the Sun, at mag. 7.8.

The path of the Sun and the planets along the ecliptic in April.

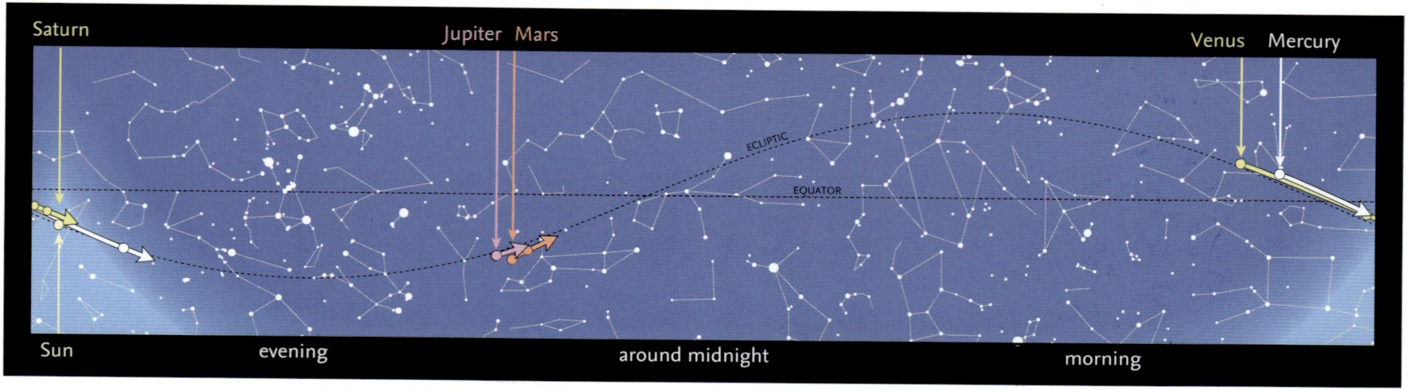

Calendar for April

01	15:07	Mars ends retrograde motion
04	06:49	Venus (mag. -3.9) 3.3°S of the Moon
06	23:51	New Moon
10	03:30	Pleiades 0.9°S of the Moon
13	03:24	Jupiter ends retrograde motion
13	22:57	First Quarter Moon
13	23:23	Pollux 4.7°N of the Moon
14	00:41	Moon at perigee = 370,001 km
15	15:01	Jupiter (mag. -2.2) 1.0°N of the Moon; an occultation will be visible from Zambia, Zimbabwe, Angola, Botswana, Namibia, South Africa and Madagascar
16	01:00	Mars (mag. -0.02) 3.3°N of the Moon
16	13:21	Regulus 1.6°N of the Moon
20	12:09	Spica 2.8°N of the Moon
20	22:27	Full Moon
23	23:16	Antares 0.1°N of the Moon; an occultation will be visible from Sub-Saharan Africa
24		Pi (π) Puppid meteor shower maximum
27	21:21	Moon at apogee = 404,170 km
28	20:18	Last Quarter Moon

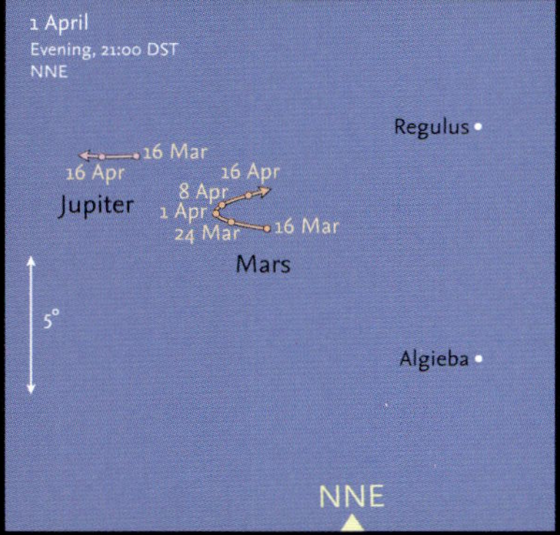

1 April • *Mars ends its retrograde motion in Leo.*

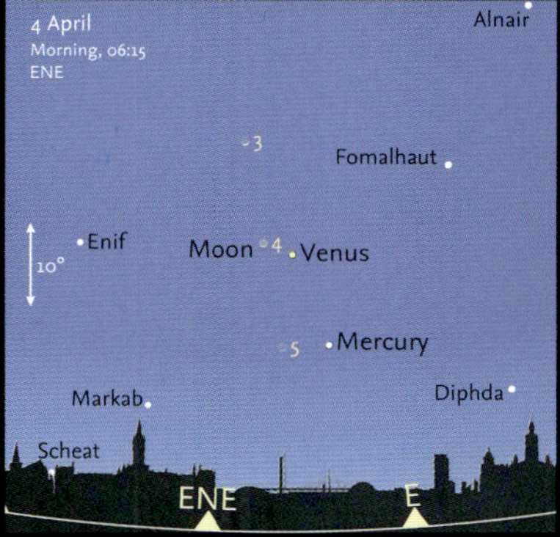

4 April • *Venus and the waning crescent Moon, with Mercury close by.*

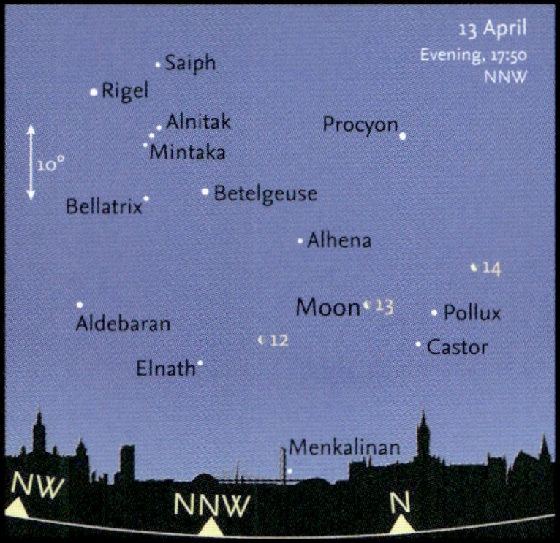

13 April • *First Quarter Moon close to Pollux in Gemini.*

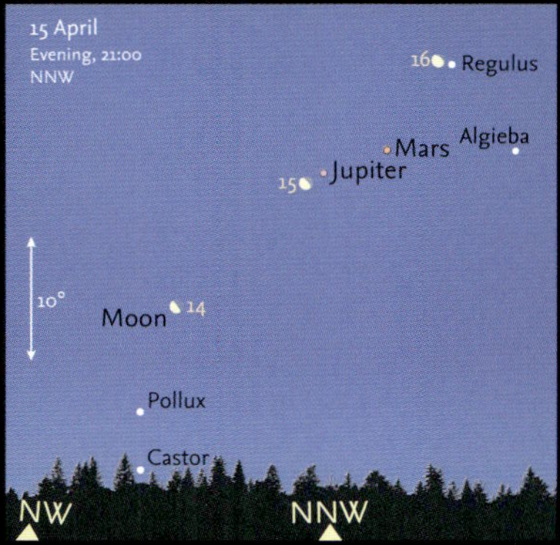

15 April • *Mars and Jupiter with Regulus and the waxing gibbous Moon.*

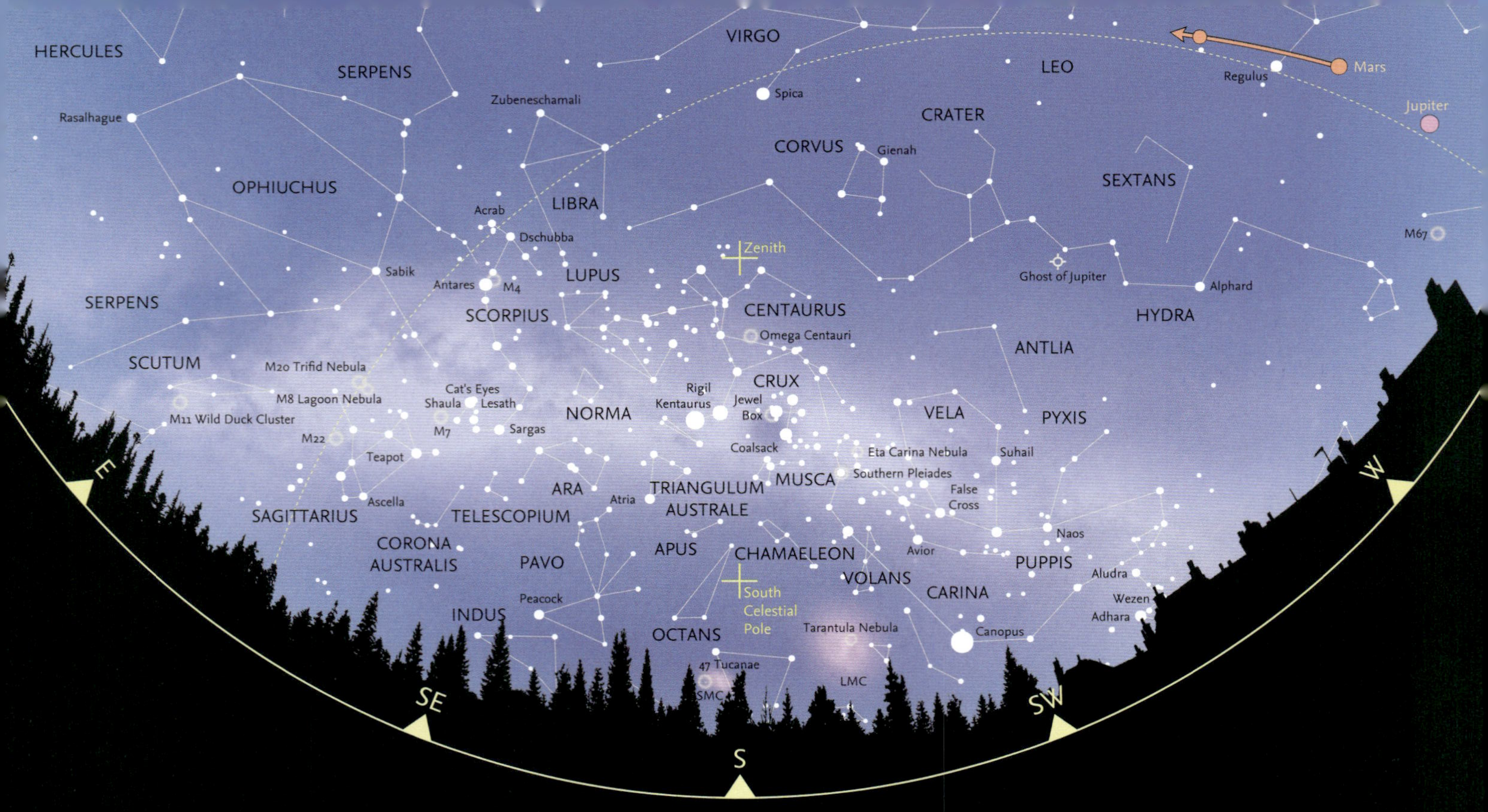

May – Looking South

1 May	23:00
15 May	22:00
1 June	21:00

May – Looking South

In the west, ***Canis Major*** has set below the horizon, and ***Canopus*** (α Carinae) and ***Puppis*** are getting rather low. The whole of ***Sagittarius*** is now clearly seen in the east, with ***Corona Australis*** beneath it and ***Scorpius*** higher above. ***Centaurus*** is fully seen high in the south, with ***Crux***, the ***Eta Carinae Nebula***, the ***Southern Pleiades*** and the ***False Cross*** with ***Vela*** along the Milky Way to the west. In the south, ***Achernar*** (α Eridani) is skimming the southern horizon, and the ***Small Magellanic Cloud*** (SMC) is beginning to rise higher in the sky, unlike the ***Large Magellanic Cloud*** (LMC), which is now lower. ***Pavo*** and ***Peacock*** (α Pavonis) are now much higher and even the faint constellations of ***Tucana*** and ***Indus*** are visible between Pavo and the southwestern horizon.

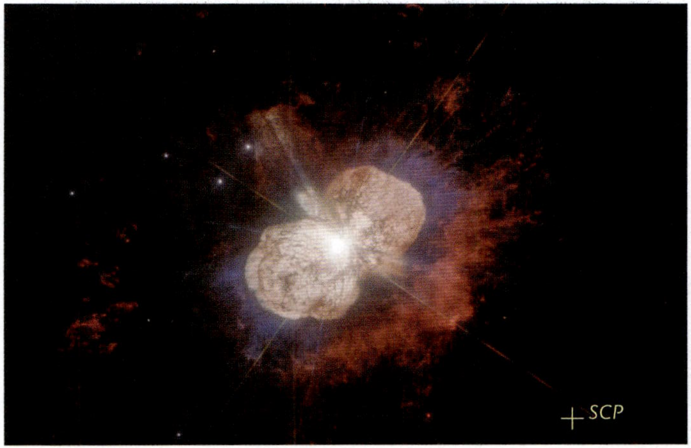

The star Eta Carinae, a blue hypergiant star within the Eta Carinae (Carina) Nebula.

Meteors

The ***η-Aquariids*** are one of the two meteor showers associated with Comet 1P/Halley, Halley's Comet, last seen in 1986 and not due until 2061. The other shower arising from the cometary debris is the ***Orionids***, in October. The radiant is near the celestial equator, close to the Water Jar in ***Aquarius***, above the horizon a few hours after midnight, also the best time to see the shower as the Earth turns into the debris. However, meteors may still be seen in the eastern sky even when the radiant is below the horizon. There is a radiant map for the η-Aquariids on page 30.

The peak of the shower in the early hours of 6 May coincides with a New Moon, meaning optimum conditions for spotting faint meteors. Maximum hourly rate is about 40 per hour, and a quarter of the meteors leave persistent trains, which can last for several seconds to minutes.

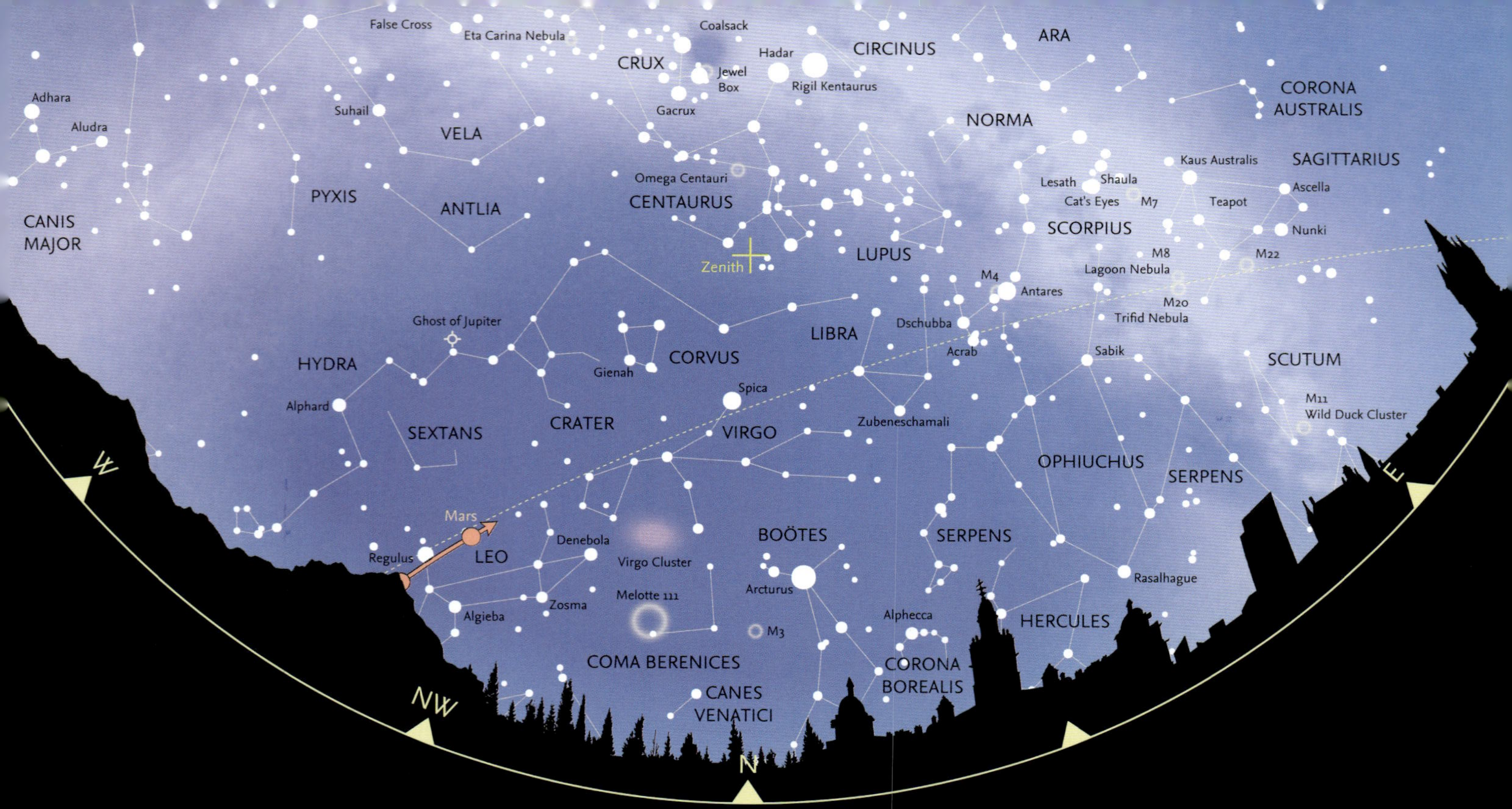

May – Looking North

1 May	23:00
15 May	22:00
1 June	21:00

May – Looking North

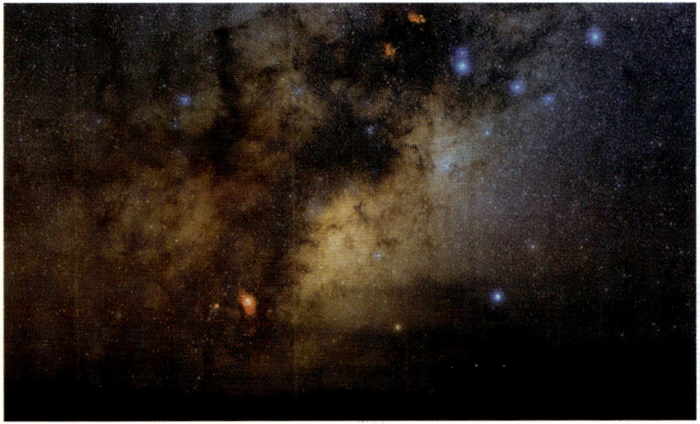

The constellation Scorpius.

Boötes is almost due north, with brilliant, slightly orange-coloured ***Arcturus*** extremely prominent. The distinctive circlet of ***Corona Borealis*** is clearly visible to its east. The brightest star (α Coronae Borealis) is known as ***Alphecca***. ***Hercules*** is rising in the east and becomes clearly visible later in the night. To the west of Boötes is ***Coma Berenices***, with the cluster ***Melotte 111*** (see page 55) and, above it, the Virgo Cluster of galaxies. ***Virgo*** contains the nearest large cluster of galaxies, which is the centre of the Local Supercluster, of which the Milky Way galaxy forms part. The Virgo Cluster contains some 2,000 galaxies, the brightest of which are visible in amateur telescopes.

The large constellation of ***Ophiuchus*** (which actually crosses the ecliptic and is thus the 'thirteenth' zodiacal constellation) is climbing into the eastern sky. Before the constellation boundaries were formally adopted by the International Astronomical Union in 1930, the southern region of Ophiuchus was regarded as forming part of the constellation of ***Scorpius***, which had been part of the zodiac since antiquity.

Early in the night, Virgo, with ***Spica*** (α Virginis), lies due north, with the rather faint constellation of ***Libra*** to its east. Farther along the ecliptic are Scorpius and brilliant, reddish ***Antares*** (α Scorpii). In the west, the constellation of ***Leo*** is still readily visible, and both ***Regulus*** and ***Denebola*** (α and β Leonis, respectively) are prominent.

The Moon's phases for May 2027

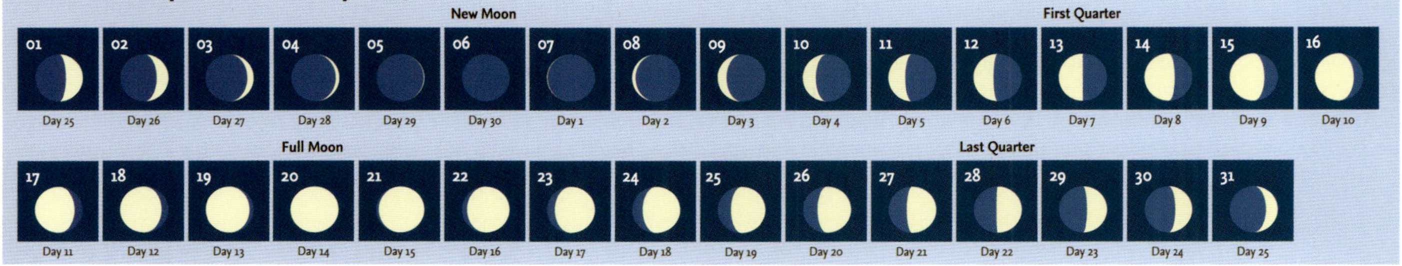

May – Moon and Planets

The Moon

The month starts with a waning crescent Moon culminating in a New Moon on 6 May. On 11 May ***Pollux*** is placed 4.9°N of the Moon, the following night ***Jupiter*** (mag. -2.0) appears 1.3°N of the Moon, high in the sky after sunset. The First Quarter Moon, on 13 May, is 1.9°S of ***Regulus*** and 3.3°S of ***Mars*** (mag. 0.4). ***Spica*** sits 2.9°N of the waxing gibbous Moon on 17 May. On 20 May the Moon is Full, the day after ***Antares*** and the Moon are less than a tenth of a degree apart a few hours before sunrise. The Moon reaches Last Quarter phase on 28 May.

The Planets

Mercury appears just after sunset and reaches eastern elongation on 28 May, it dims from mag. -2.1 to 0.7 through May. ***Venus*** pops up just above the horizon before sunrise at mag. -3.9. It approaches ***Saturn*** (mag. 0.7) on 7 May, passing 0.6°N of the ringed gas giant. ***Mars*** stays in ***Leo*** after sunset at mag. 0.3, dimming to mag. 0.7 as the month progresses. On 13 May Mars (mag. 0.4) can be found 3.3°N of the First Quarter Moon and the day after it will be 1.1°N of ***Regulus***. ***Jupiter*** lies in ***Cancer*** until 10 May, after which it moves into Leo. It is visible most of the night, dimming from mag. -2.1 to -1.9 at the end of May. Jupiter (mag. -2.0) lies 1.3°N of the waxing crescent Moon on 12 May. Saturn lies in ***Pisces***, rising just before the Sun at mag. 0.7. ***Uranus*** is visible in ***Taurus*** in the early evening, at mag. 5.8. ***Neptune*** is in Pisces rising just before the Sun, at mag. 7.8.

The path of the Sun and the planets along the ecliptic in May.

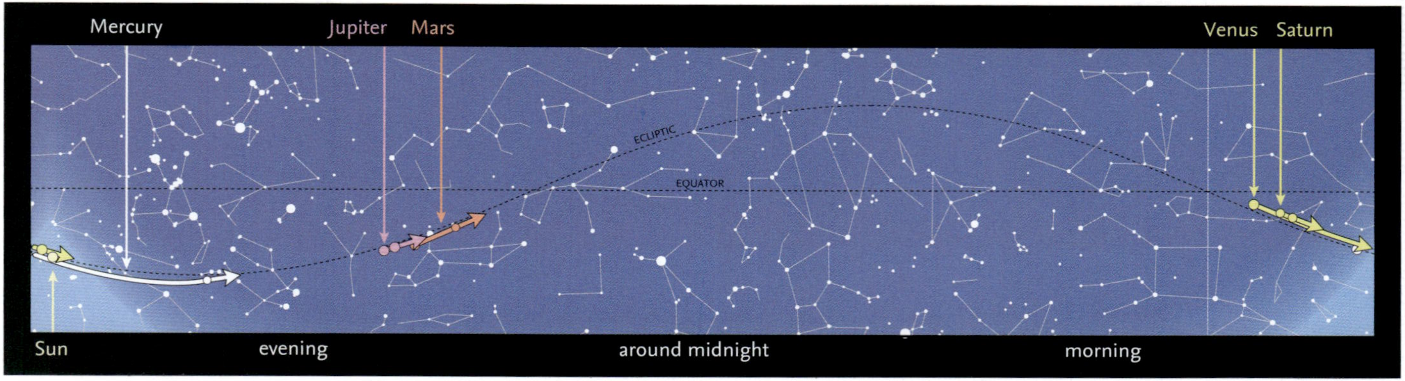

Calendar for May

05–06		Eta (η) Aquariid meteor shower maximum
06	10:58	New Moon
07	16:00	Venus (mag. -3.9) 0.6°N of Saturn (mag. 0.7)
09	20:07	Moon at perigee = 366,635 km
11	05:06	Pollux 4.9°N of the Moon
12	22:49	Jupiter (mag. -2.0) 1.3°N of the Moon
13	04:44	First Quarter Moon
13	18:57	Regulus 1.9°N of the Moon
13	19:34	Mars (mag. 0.4) 3.3°N of the Moon
14	10:58	Mars (mag. 0.4) 1.1°N of Regulus
17	19:16	Spica 2.9°N of the Moon
20	10:59	Full Moon
21	07:08	Antares 0.0°S of the Moon; an occultation will be visible from Brazil, Peru, northern Argentina and Bolivia
25	15:13	Moon at apogee = 404,801 km
28	13:58	Last Quarter Moon
28	14:13	Mercury at greatest elongation (22.9°E, mag. 0.4)

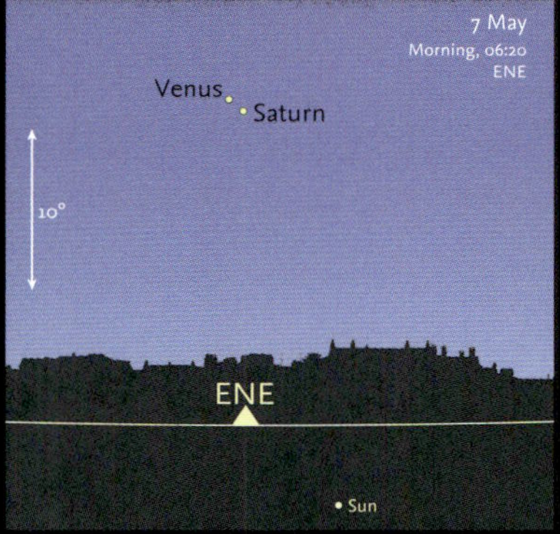

7 May • *Venus and Saturn low in the east.*

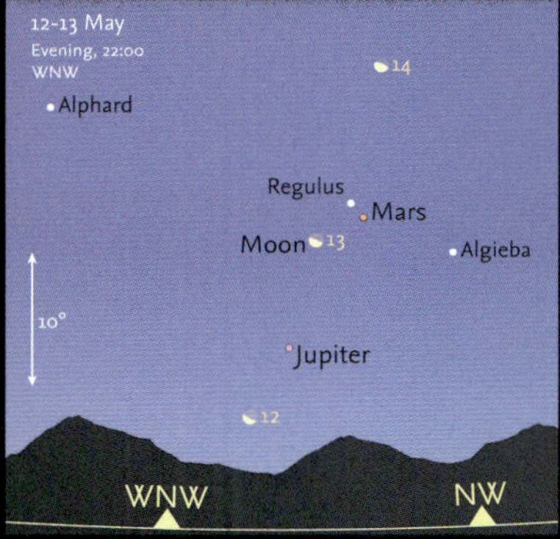

12–13 May • *Waxing crescent Moon swings by Jupiter, Mars and Regulus.*

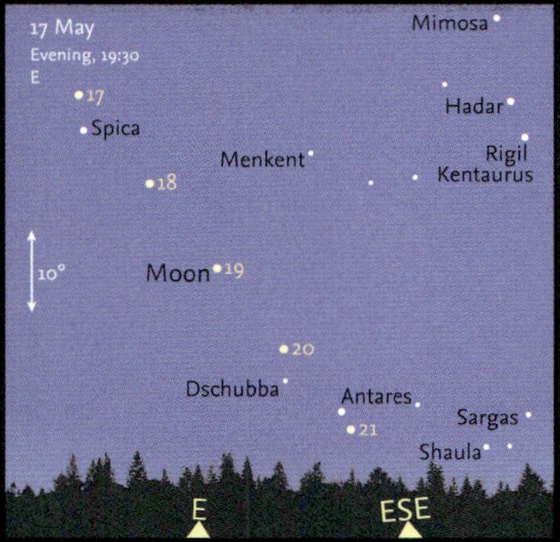

17 May • *Virgo's brightest star, Spica, close to the waxing gibbous Moon.*

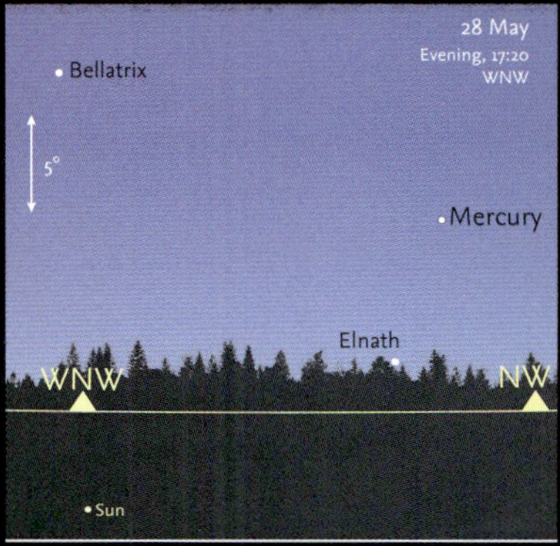

28 May • *Mercury at greatest eastern elongation.*

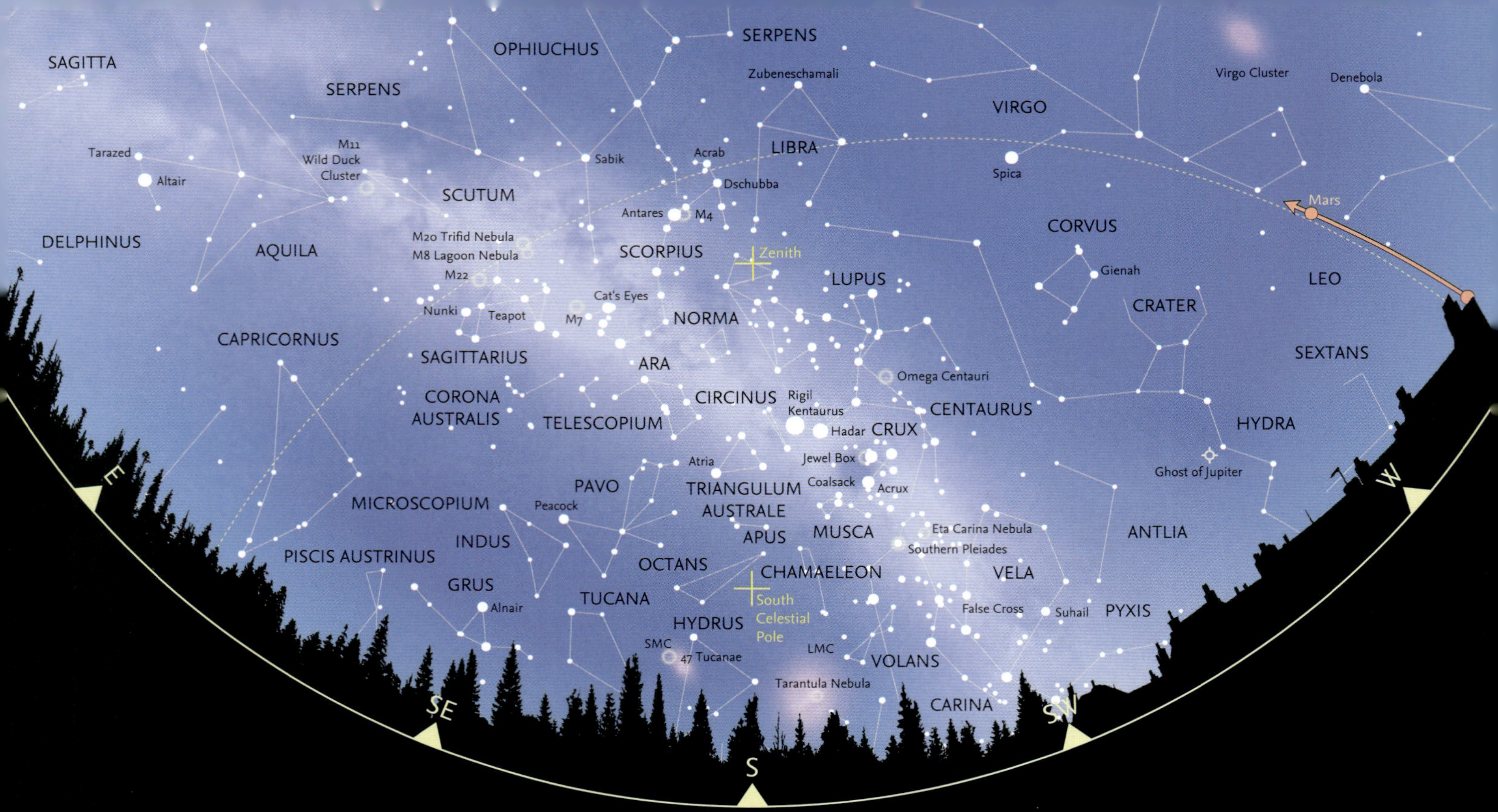

June – Looking South

1 June	23:00
15 June	22:00
1 July	21:00

June – Looking South

The winter solstice on 21 June brings the shortest period of daylight. Although the whole of ***Carina*** is visible, all of ***Eridanus*** (except ***Acherna***r) is hidden below the horizon. The ***Large Magellanic Cloud*** (LMC) is low, although the ***Small Magellanic Cloud*** (SMC), the globular cluster, ***47 Tucanae*** and ***Hydrus*** are now rather higher. ***Alphard*** (α Hydrae) is very low on the horizon, and the constellation of ***Sextans*** is becoming low. However, the remainder of the long constellation of ***Hydra*** is clearly seen as are the two constellations of ***Crater*** and ***Corvus*** to its north. The ***False Cross*** between Carina and ***Vela*** is beginning to descend in the west, but Vela itself is clearly visible. The whole of both ***Crux*** and ***Centaurus*** are clearly seen, as are the magnificent globular cluster, ***Omega Centauri***, and the constellation of ***Lupus***, closer to the zenith. The constellation of ***Triangulum Australe*** is on the meridian, roughly halfway between the South Celestial Pole and the zenith. Both ***Scorpius***, with brilliant, red ***Antares*** (α Scorpii) and ***Sagittarius*** are high overhead, with the faint constellation of ***Corona Australis*** visible below them. The whole of ***Capricornus*** is visible, and the constellation of ***Grus*** has now risen above the horizon, with the faint constellation of ***Indus*** between it and ***Pavo***. To the east of Grus is ***Piscis Austrinus***, although brilliant ***Fomalhaut*** (α Piscis Austrini) is only just clear of the horizon and becomes clearly visible only later in the night and later in the month.

The central region of the Milky Way, Sagittarius A.*

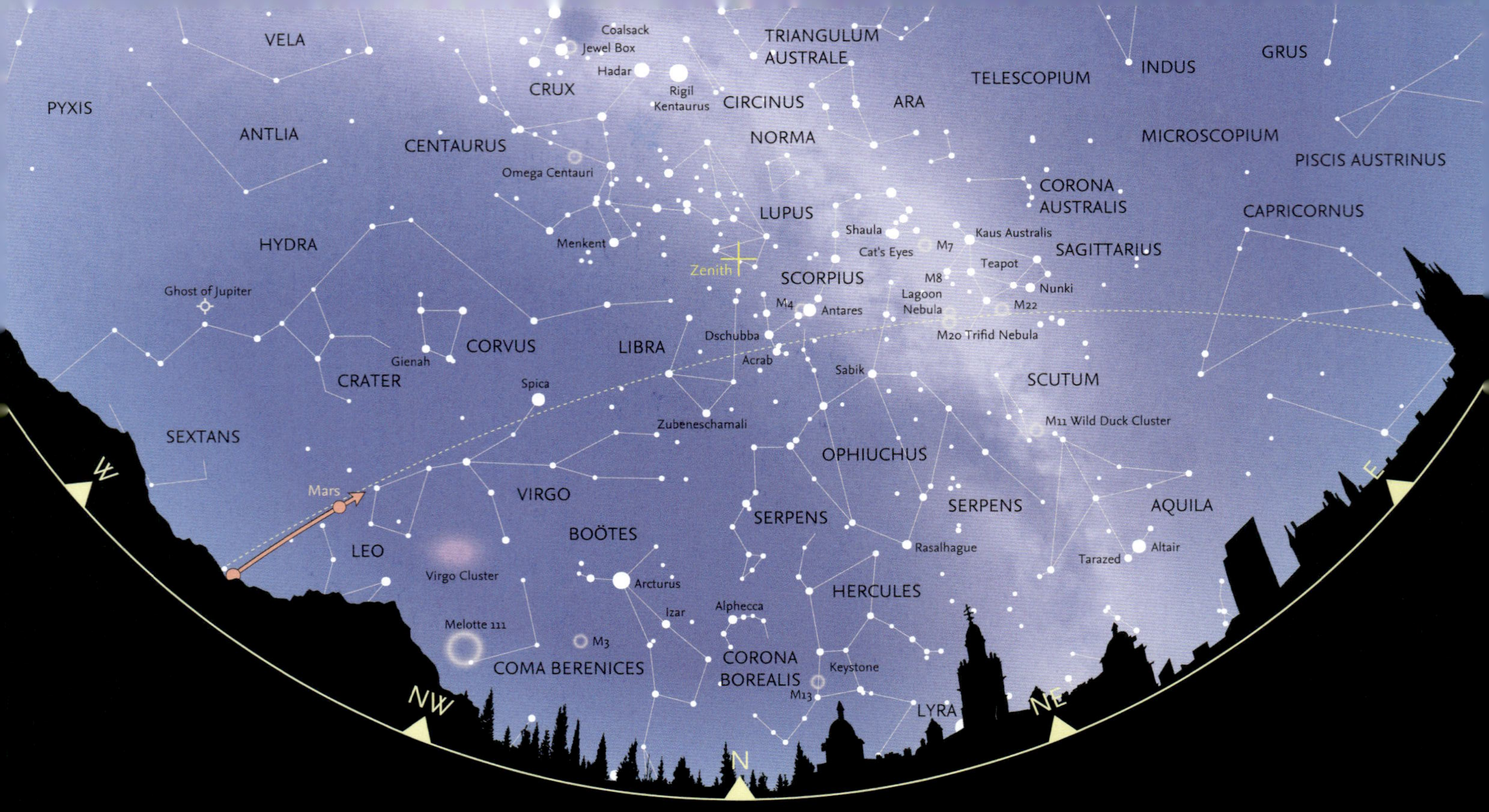

June – Looking North

1 June	23:00
15 June	22:00
1 July	21:00

June – Looking North

Vega in ***Lyra*** is just above the northeastern horizon. Closer to the meridian, the whole of ***Hercules*** is visible including the Keystone and ***M13***, a globular cluster first discovered by Astronomer Royal Edmond Halley in 1714. The small constellation of ***Corona Borealis*** is almost due north. To the west of the meridian is ***Boötes*** and ***Arcturus*** (α Boötis). Much of ***Leo*** is now below the western horizon, but ***Denebola*** (β Leonis) is still visible. Higher in the sky, the whole of ***Virgo*** is clearly visible and, still higher, not far from the zenith is the constellation of ***Libra***. To its east is ***Scorpius*** and the red supergiant ***Antares*** – the name means the 'rival of Mars'. Farther along the ecliptic, both the constellations of ***Sagittarius*** and ***Capricornus*** are completely visible. Between Hercules and Sagittarius is the large constellation of ***Ophiuchus*** (the Serpent Bearer), lying between the two halves of the constellation of ***Serpens***: ***Serpens Caput*** (Head of the Serpent) to the west and ***Serpens Cauda*** (Tail of the Serpent) to the east. The ecliptic runs across Ophiuchus and, to its east, ***Aquila*** and ***Altair*** (α Aquilae), one of the stars of the (northern) Summer Triangle.

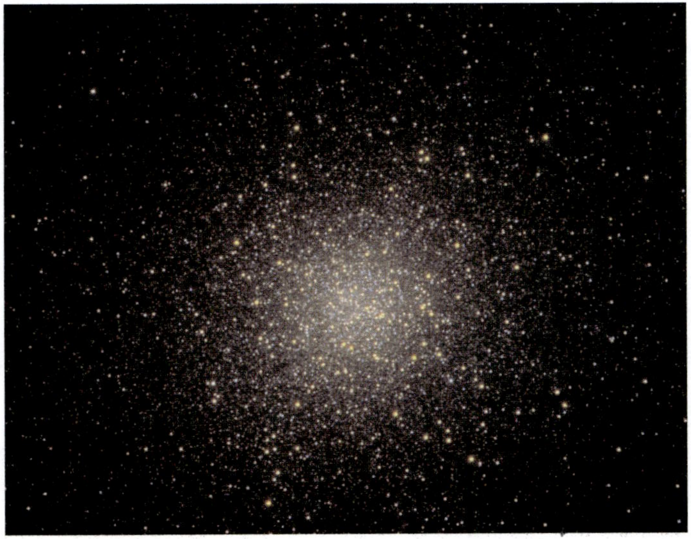

M13, the Great Globular Cluster, in Hercules.

The Moon's phases for June 2027

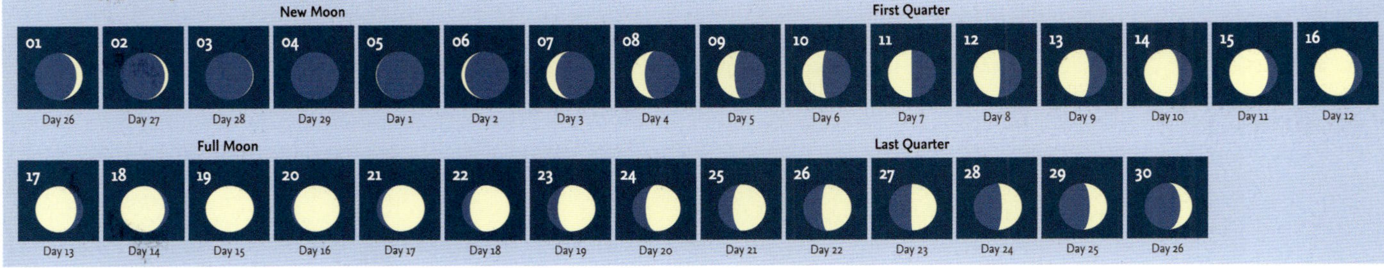

June – Moon and Planets

The Moon

A New Moon occurs on 4 June and two days later ***Mercury*** (mag. 1.4) appears 2.7°S of the waxing crescent Moon, prominent in the sky just before sunset. On 9 June ***Jupiter*** (mag. -1.9) sits 1.9°N of the Moon, the following day it's the turn of ***Mars*** (mag. 0.8) passing 3.9°N of the Moon and ***Regulus*** lies 2.2°N of the Moon. The First Quarter Moon is present on 11 June. On 14 June ***Spica*** lies 3.0°N of the waxing gibbous Moon, three days later ***Antares*** passes less than a tenth of a degree of the Moon, low in the southeast after sunset. The Moon is Full on 19 June and Last Quarter Moon arises on 27 June.

The Planets

Mercury appears just after sunset and switches to the early morning after 23 June. Its magnitude fluctuates, dimming from mag. 0.6 to 5.9 and then brightening to mag. 3.6 at the end of the month. On 6 June Mercury (mag. 1.4) sits 2.7°S of the thin waxing crescent Moon low in the sky. ***Venus*** is visible at dawn at mag. -3.9. On 16 June Venus passes 4.7°N of ***Aldebaran*** in ***Taurus***, a challenging sight through the glow of the sunrise. ***Mars*** stays in ***Leo***, visible after sunset and starting the month at mag. 0.8, dimming to mag. 1.1 at the end of June. On 10 June Mars (mag. 0.8) will appear 3.9°N of the bright waxing crescent Moon. ***Jupiter*** stays in ***Leo*** with Mars, visible until midnight at mag. -1.9, dimming slightly to -1.8 as the month progresses. Jupiter (mag. -1.9) will appear 1.9°N of the waxing crescent Moon on 9 June. ***Saturn*** lies in ***Pisces***, rising an hour or so before the Sun at mag. 0.7, brightening slightly to mag. 0.6 towards the end of the month. ***Uranus*** sits in ***Taurus***, setting with the Sun, at mag. 5.8. ***Neptune*** is in Pisces, rising at dawn, at mag. 7.8.

The path of the Sun and the planets along the ecliptic in June.

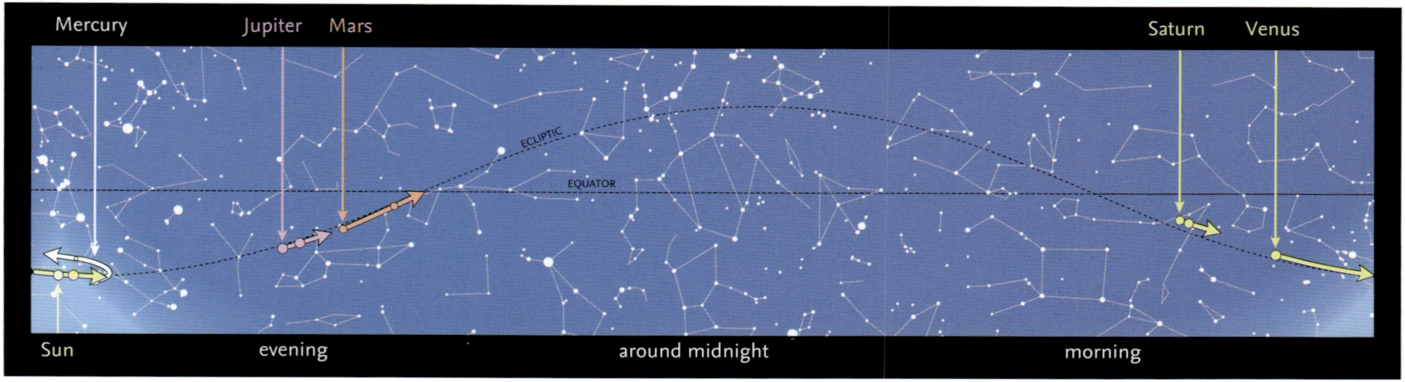

Calendar for June

04	19:40	New Moon
06	06:29	Mercury (mag. 1.4) 2.7°S of the Moon
06	14:54	Moon at perigee = 361,703 km
09	10:54	Jupiter (mag. -1.9) 1.9°N of the Moon
10	01:01	Regulus 2.2°N of the Moon
10	22:45	Mars (mag. 0.8) 3.9°N of the Moon
11	10:56	First Quarter Moon
14	01:01	Spica 3.0°N of the Moon
16	10:11	Venus (mag. -3.9) 4.7°N of Aldebaran
17	13:46	Antares 0.0°S of the Moon; an occultation will be visible from Australia, western Indonesia, eastern Papua New Guinea
19	00:44	Full Moon
21	14:11	Summer Solstice
22	05:08	Moon at apogee = 405,698 km

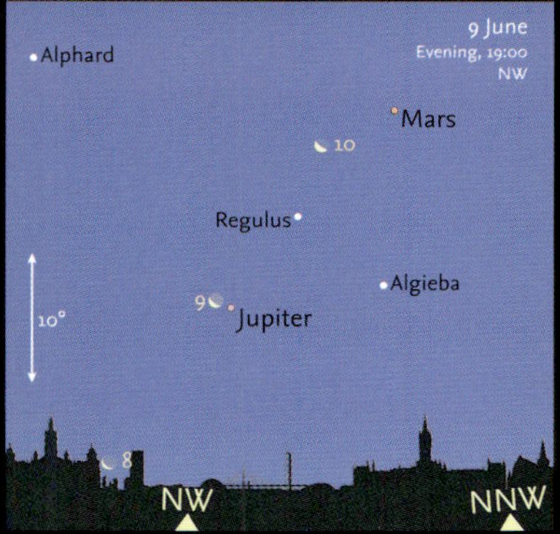

9 June • *The waxing crescent Moon with Jupiter and Mars in Leo.*

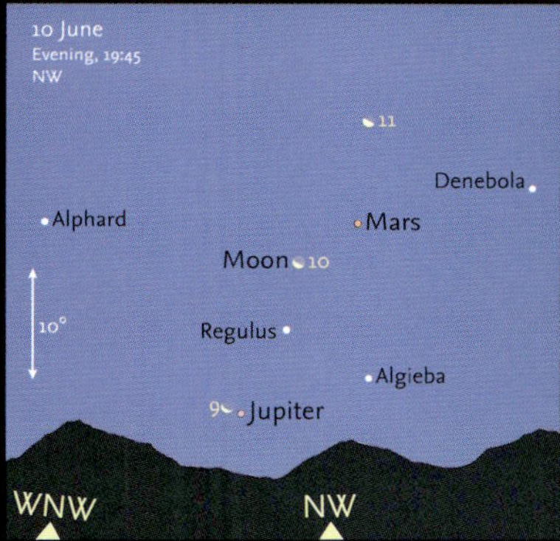

10 June • *The Moon next to Mars just before First Quarter.*

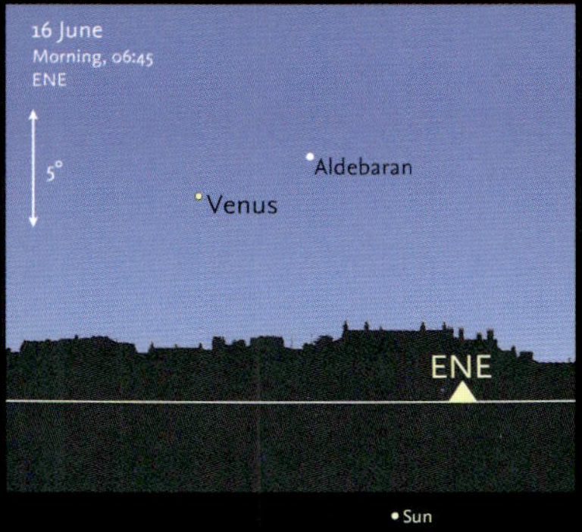

16 June • *Venus in Taurus, close to Aldebaran.*

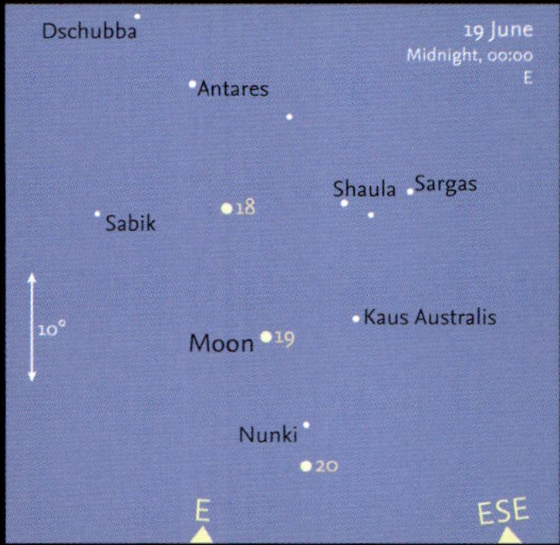

19 June • *Full Moon in Sagittarius.*

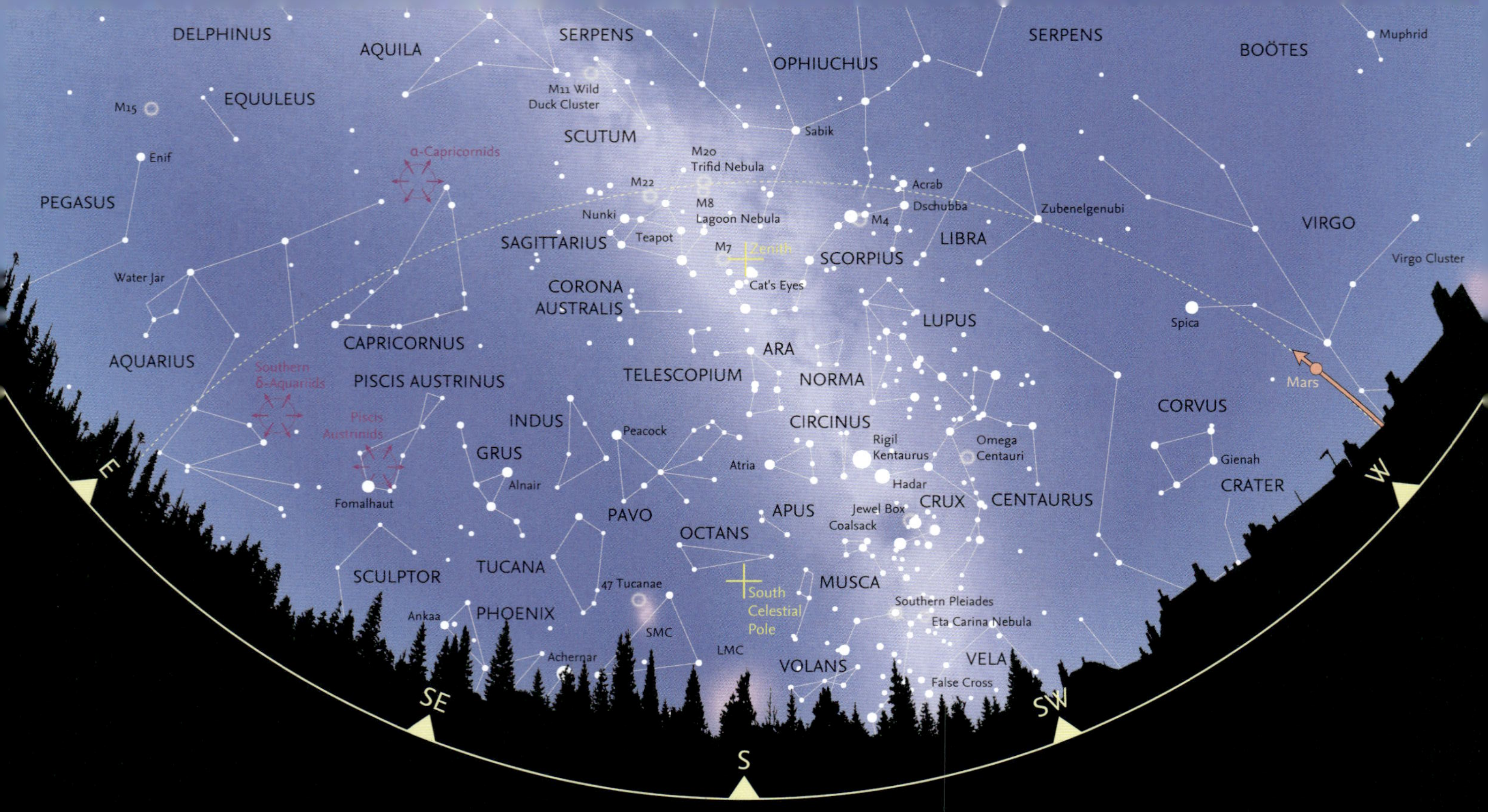

July – Looking South

1 July	23:00
15 July	22:00
1 August	21:00

July – Looking South

Although the ***Large Magellanic Cloud*** (LMC) is almost due south, it is very low. Slightly farther west, the ***False Cross*** is nearing the horizon. Some of ***Hydra*** remains visible, but the constellation of ***Crater*** is becoming very low. ***Corvus*** is still easy to see. The zodiacal constellation of ***Virgo*** will soon be disappearing in the west. ***Crux***, ***Centaurus*** and ***Lupus*** are still clearly seen, high in the sky, as is ***Scorpius***, the tail of which is near the zenith. ***Achernar*** (α Eridani), ***Hydrus*** and the ***Small Magellanic Cloud*** (SMC) are now higher above the horizon and easier to observe. The whole of the constellation of ***Phoenix*** is also now clear of the horizon. Above it are ***Tucana*** and, halfway to the zenith, the constellation of ***Pavo***. ***Grus*** and ***Pisces Austrinus*** (with brilliant ***Fomalhaut***) are now fully visible. Higher in the sky are the zodiacal constellations of ***Capricornus*** and ***Aquarius***, and even the westernmost portion of ***Pisces*** is rising above the horizon.

Meteors

July brings three meteor showers; there are two minor radiants active in the constellations of ***Capricornus*** and ***Aquarius***. The first shower, the ***α-Capricornids***, active from 3 July–15 August (peaking 30 July), has a maximum rate of about five per hour; however, it does often produce very bright fireballs. The parent body is Comet 169P/NEAT. The most prominent shower is probably that of the ***Southern δ-Aquariids***, which are active from around 12 July–23 August, with a peak on 30 July, although even then the rate is unlikely to reach 25 meteors per hour. In this case, the parent body may be Comet 96P/Machholz. This year, both shower maxima occur when the Moon is a thin waning crescent, meaning there will be very little moonlight present, the Moon rising after midnight. The ***Piscis Austrinids*** begin on 15 July and continue until 10 August. Maximum is on 28 July, but the rate is only about five per hour. The ***Perseids*** begin on 17 July and peak on 12–13 August.

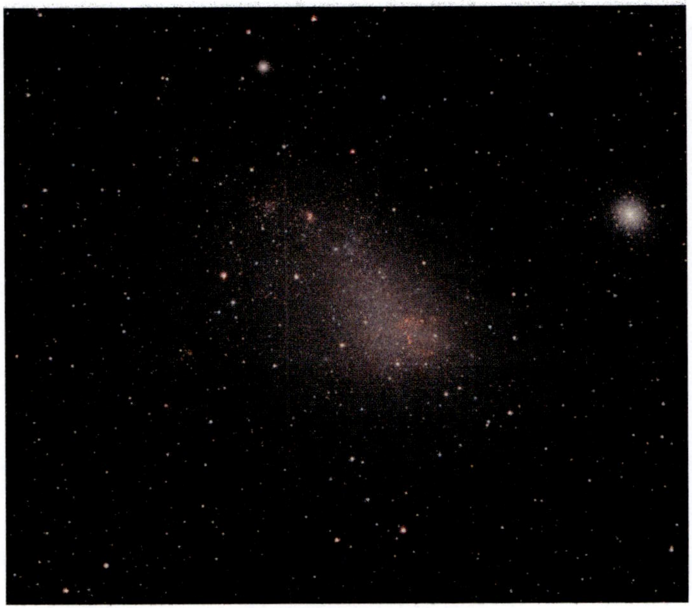

The Small Magellanic Cloud, a dwarf galaxy orbiting the Milky Way.

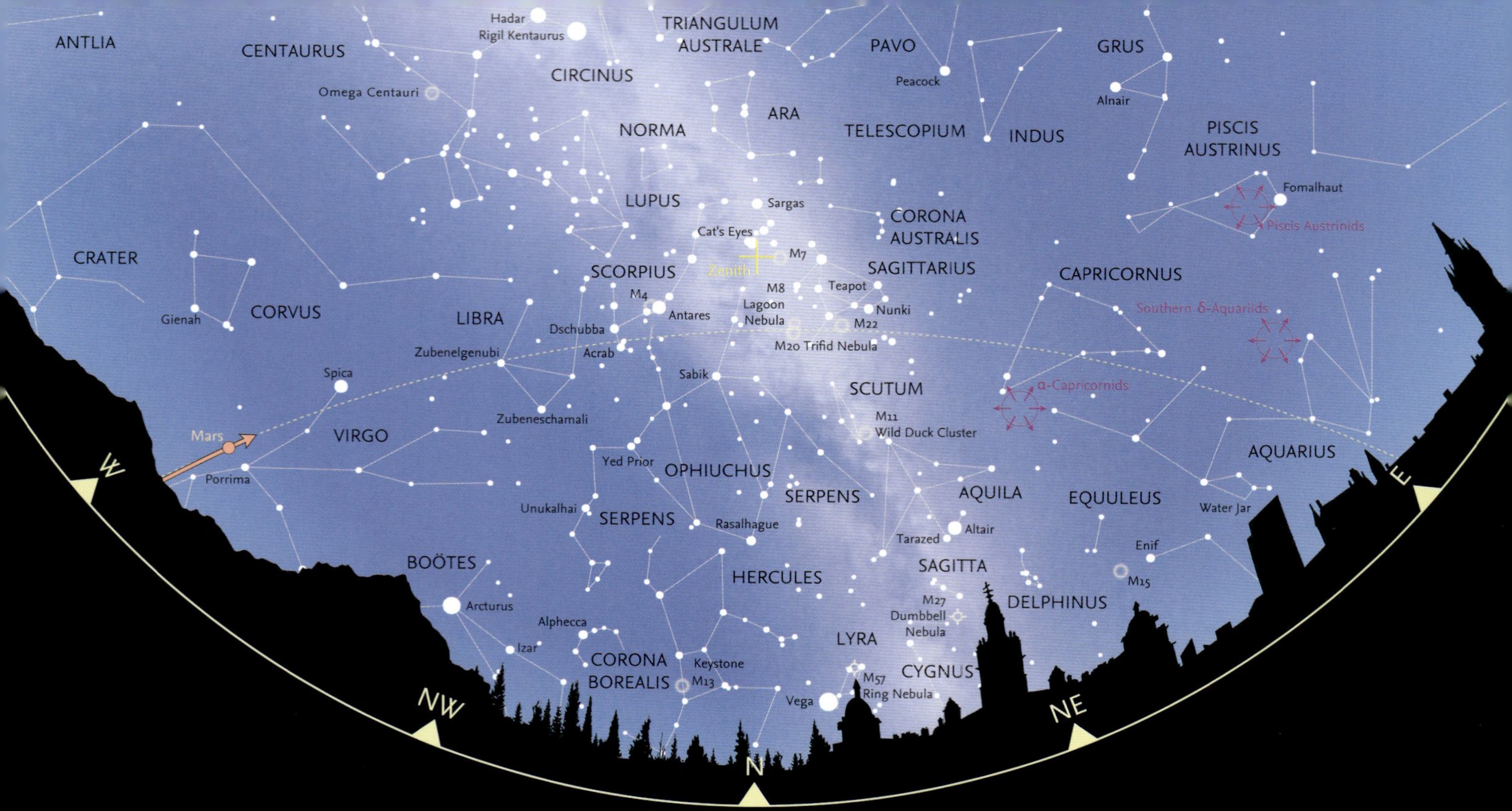

July – Looking North

1 July	23:00
15 July	22:00
1 August	21:00

July – Looking North

The constellations of ***Hercules*** and ***Lyra*** are on either side of the meridian. To the east, ***Deneb*** (α Cygni) is skimming the horizon, but nearly all the rest of the constellation is easily seen, as is ***Aquila*** with ***Altair*** (α Aquilae). Between ***Cygnus*** and Aquila lies the small constellation of ***Sagitta***, with ***M27*** (the Dumbbell Nebula, faintly visible through binoculars or telescopes as inverted triangles). Even farther east, both the zodiacal constellations of ***Aquarius*** and ***Capricornus*** are visible. Above Aquila, farther along the ***Great Dark Rift*** – dust clouds that hide the more distant stars – is the small constellation of ***Scutum***, with ***M11***, the Wild Duck Cluster. To its east is the tiny constellation of ***Delphinus.*** Still farther along the Great Rift lies the centre of the Milky Way Galaxy and, near it, two emission nebulae: ***M8*** (the Lagoon Nebula) and ***M20*** (the Trifid Nebula). In the western sky, the constellation of ***Boötes*** and ***Arcturus*** (α Boötis) are beginning to approach the horizon, but ***Corona Borealis*** is still clearly seen. Even farther west, the whole of ***Virgo*** is visible, with ***Libra*** above it. The large constellation of ***Ophiuchus*** and the two halves of ***Serpens*** lie between Hercules and the zenith. ***Scorpius*** is draped around the zenith.

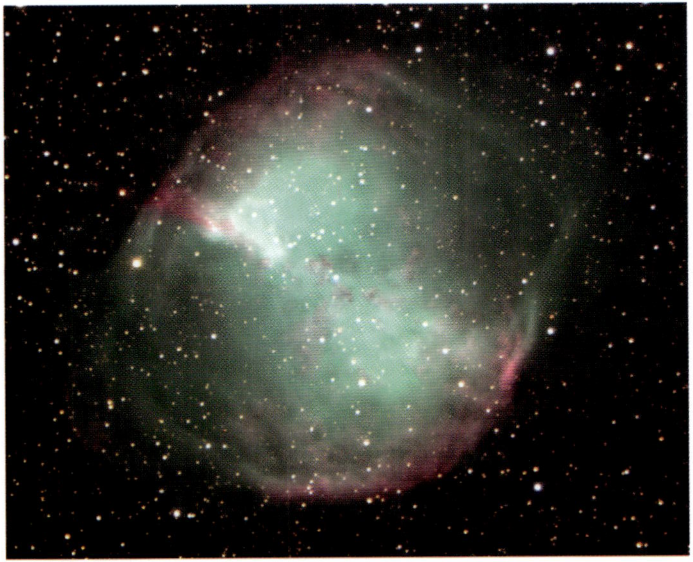

The Dumbbell Nebula, a planetary nebula – the remnant of a dying star.

The Moon's phases for July 2027

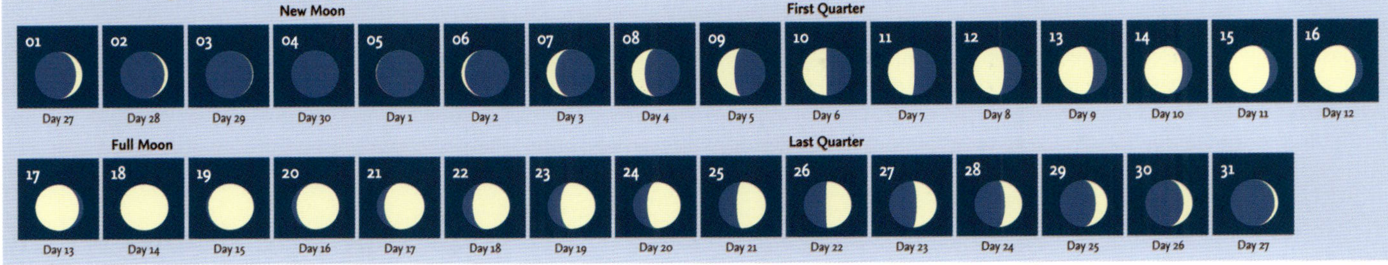

July – Moon and Planets

The Moon

The first day of July brings the ***Pleiades*** and the Moon together in the sky, with the star cluster 0.9°S of the waning crescent Moon, an hour before sunrise. The Moon becomes New on 4 July, on 7 July the waxing crescent Moon sits 2.5°S of the gas giant ***Jupiter*** (mag. -1.8), with ***Regulus*** in ***Leo*** lying 2.3°N, all three dip below the horizon a few hours before midnight. The day before the First Quarter Moon on 10 July, ***Mars*** (mag. 1.1) passes 4.7°N of the Moon and on 11 July ***Spica*** is 3.2°N of the waxing gibbous Moon. On 14 July the 'rival of Mars', ***Antares*** is positioned 0.1°N of the waxing gibbous Moon. The Moon is Full on 18 July and on 26 July the Moon reaches Last Quarter. On 28 July the beautiful ***Pleiades*** are back close to the waning crescent Moon, situated only 1.0°S.

The Planets

Mercury appears in the sky at dawn, starting the month at mag. 3.3 and brightening to mag. -1.3. It reaches western elongation on 16 July. ***Venus*** continues to rise in the early hours at mag. -3.9. ***Mars*** continues in ***Leo*** until 4 July, after which it progresses into ***Virgo***, visible after sunset until just after midnight. It starts the month at mag. 1.1, gradually dimming to mag. 1.2. ***Jupiter*** stays in ***Leo*** with Mars, setting a few hours before midnight at mag. -1.8, dimming slightly to -1.7 as the month progresses. It lies 0.4°N of ***Regulus*** on 25 July at mag. -1.7. ***Saturn*** lies in ***Pisces***, appearing above the horizon a few hours before sunrise, starting the month at mag. 0.6 and slowly brightening to mag. 0.5. ***Uranus*** sits in ***Taurus***, climbing above the horizon an hour before sunrise, at mag. 5.8. ***Neptune*** is in Pisces rising shortly after midnight, gradually brightening from mag. 7.8 to mag. 7.7. It enters retrograde motion on 10 July.

The path of the Sun and the planets along the ecliptic in July.

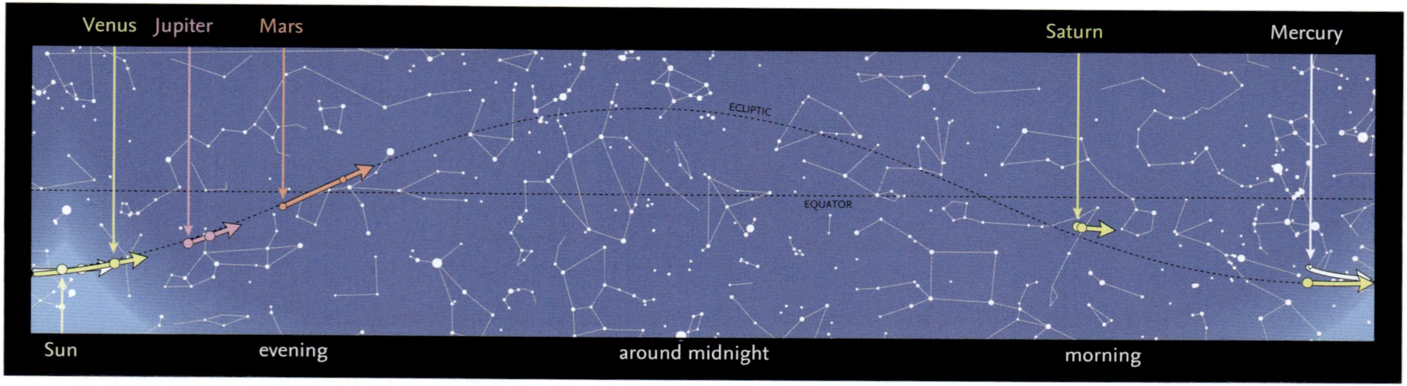

Calendar for July

01	07:04	Pleiades 0.9°S of the Moon
04	03:02	New Moon
04	20:55	Moon at perigee = 358,284 km
05	05:05	Earth at aphelion (152,100,643 km = 1.01673 AU)
07	03:14	Jupiter (mag. -1.8) 2.5°N of the Moon
07	09:01	Regulus 2.3°N of the Moon
09	07:52	Mars (mag. 1.1) 4.7°N of the Moon
10	00:30	Neptune enters retrograde motion
10	18:39	First Quarter Moon
11	06:50	Spica 3.2°N of the Moon
14	19:34	Antares 0.1°N of the Moon; an occultation will be visible from Sub-Saharan Africa
16	03:31	Mercury at greatest elongation (20.7°W, mag. 0.4)
18	15:45	Full Moon
19	11:52	Moon at apogee = 406,217 km
25	06:23	Jupiter (mag. -1.7) 0.4°N of Regulus
26	16:55	Last Quarter Moon
28	16:20	Pleiades 1.0°S of the Moon
29		Piscis Austrinid meteor shower maximum
30		Delta (δ) Aquariid meteor shower maximum
30		Alpha (α) Capricornid meteor shower maximum

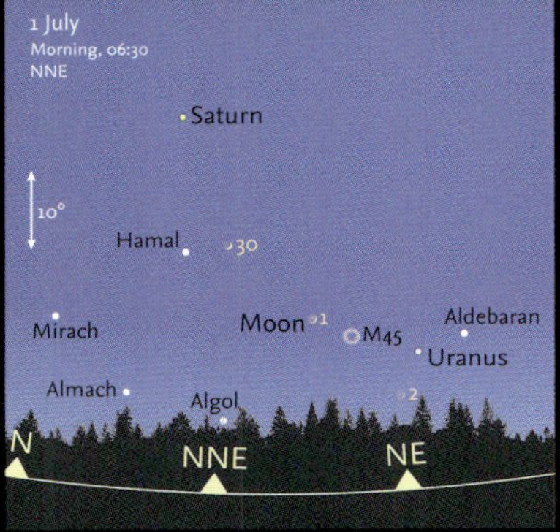

1 July • *The waning crescent Moon lies next to the Pleiades (M45). Uranus and Saturn are nearby.*

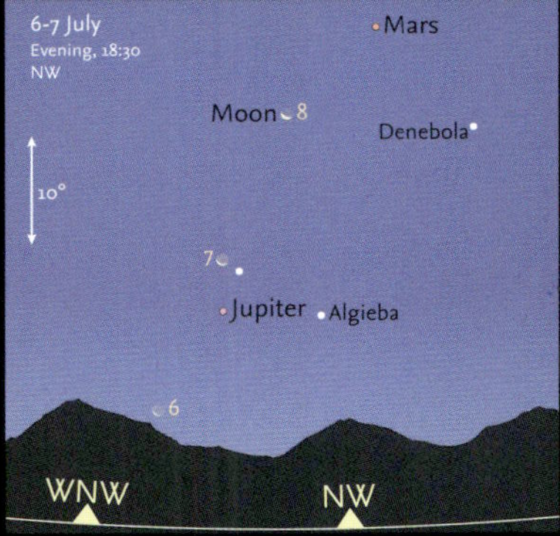

6–7 July • *Waxing crescent Moon passes by Jupiter.*

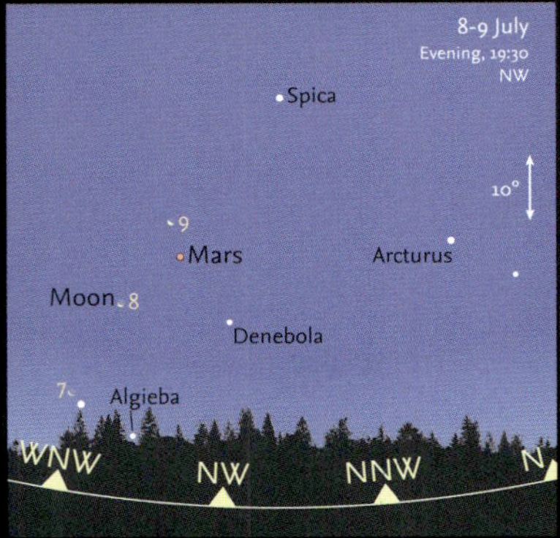

8–9 July • *The Moon swings past Mars.*

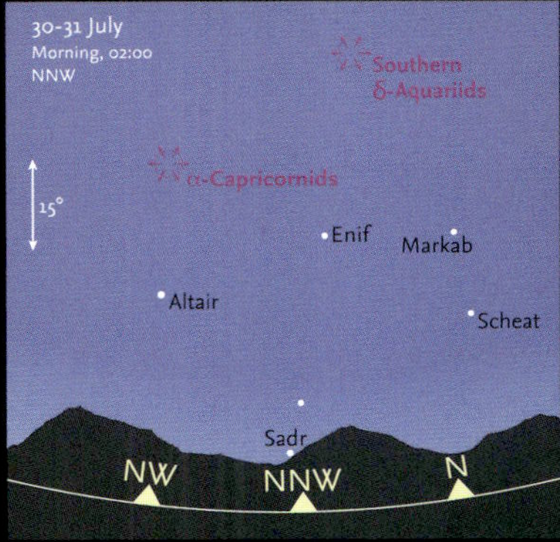

30–31 July • *The peak of the Southern δ-Aquariids.*

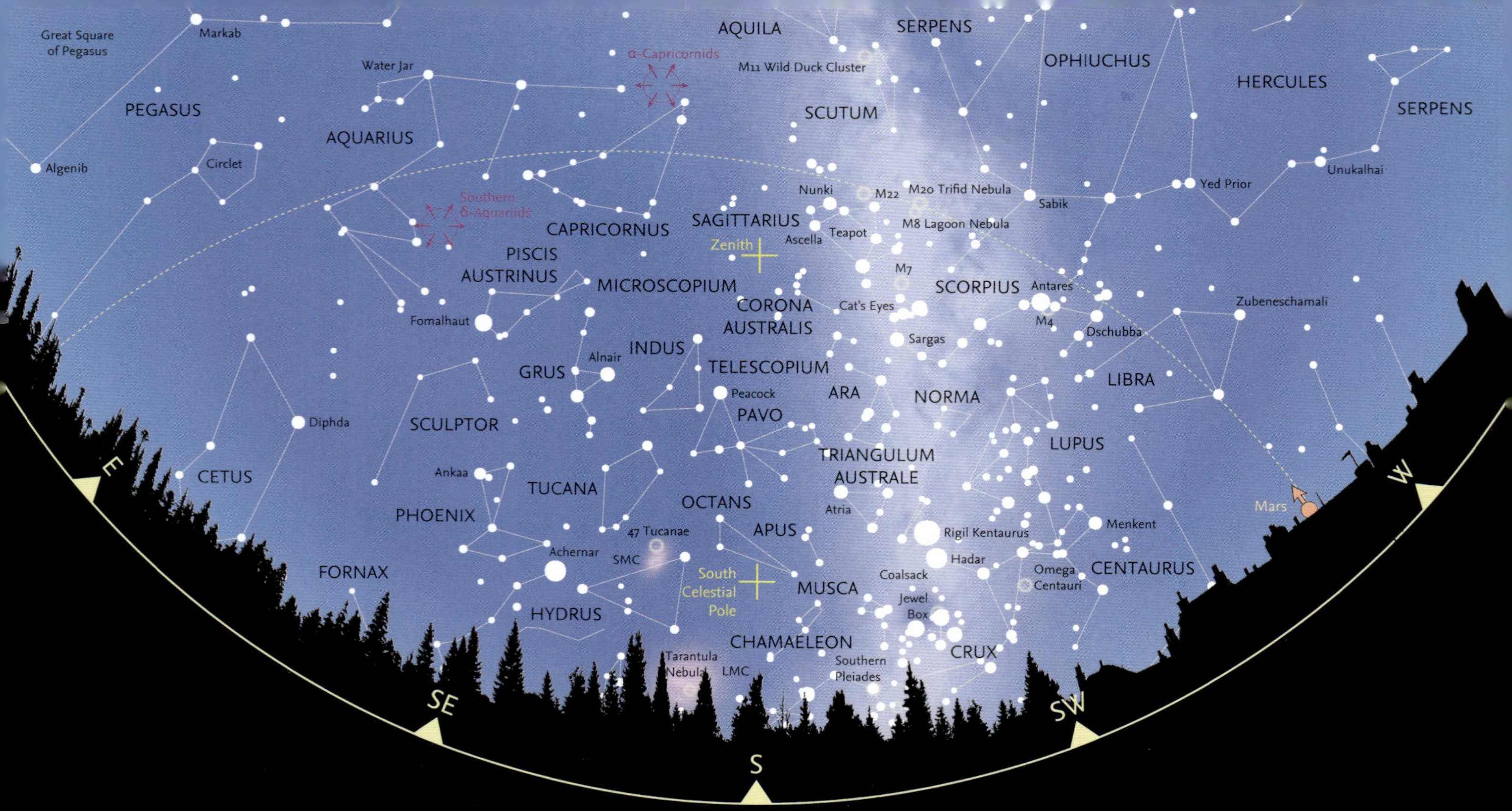

August – Looking South

1 August	23:00
15 August	22:00
1 September	21:00

August – Looking South

Three constellations: ***Volans***, ***Chamaeleon*** and ***Octans*** are on the meridian, with ***Pavo*** (the Peacock) higher towards the zenith. Much of ***Carina*** is below the horizon, although the ***Eta Carinae Nebula*** and the ***Southern Pleiades*** are still visible. The ***Large Magellanic Cloud*** (LMC), a dwarf galaxy orbiting the Milky Way, is slightly higher in the sky. ***Crux*** is now lower, but the whole of ***Centaurus*** and ***Lupus*** remains visible. Much of ***Virgo*** has set in the west and ***Libra*** is following it down towards the horizon. ***Scorpius*** and ***Sagittarius*** are still visible high in the sky. ***Achernar***, the ***Small Magellanic Cloud*** (SMC) and ***Hydrus*** are now well clear of the horizon, but below them are more small constellations: ***Dorado***, ***Reticulum*** and ***Horologium***. More of the river of ***Eridanus*** is visible, together with parts of ***Pictor*** and ***Fornax***. Much of ***Cetus*** has risen and the whole of the western arm of ***Pisces*** is now clearly seen. ***Sculptor***, ***Grus*** and ***Piscis Austrinus*** lie halfway between the eastern horizon and the zenith, with faint eighteenth-century constellation ***Microscopium*** closer to the actual zenith.

Meteors

The ***Piscis Austrinid*** shower continues until about 10 August, after maximum on 28 July. The ***Southern δ-Aquariids*** reach maximum of up to 25 meteors per hour on 30 July. The ***Perseids*** peak on 12–13 August, when the rate may reach as high as 100 meteors per hour. In 2027, there is a bright waxing gibbous at the maximum, so the moonlight will interfere with observations before midnight – meteors are generally best seen after this time and skies will be darker for this shower. The Perseids are debris from Comet 109P/Swift -Tuttle (the Great Comet of 1862). Perseid meteors are fast and many of the brighter ones leave persistent trains. Some bright fireballs also occur during the shower, arising from larger chunks of cometary debris. The ***α-Aurigid*** shower begins on 28 August and is a short shower, lasting until 5 September and reaching maximum on 1 September.

The Perseid meteor shower, known for its bright fireballs.

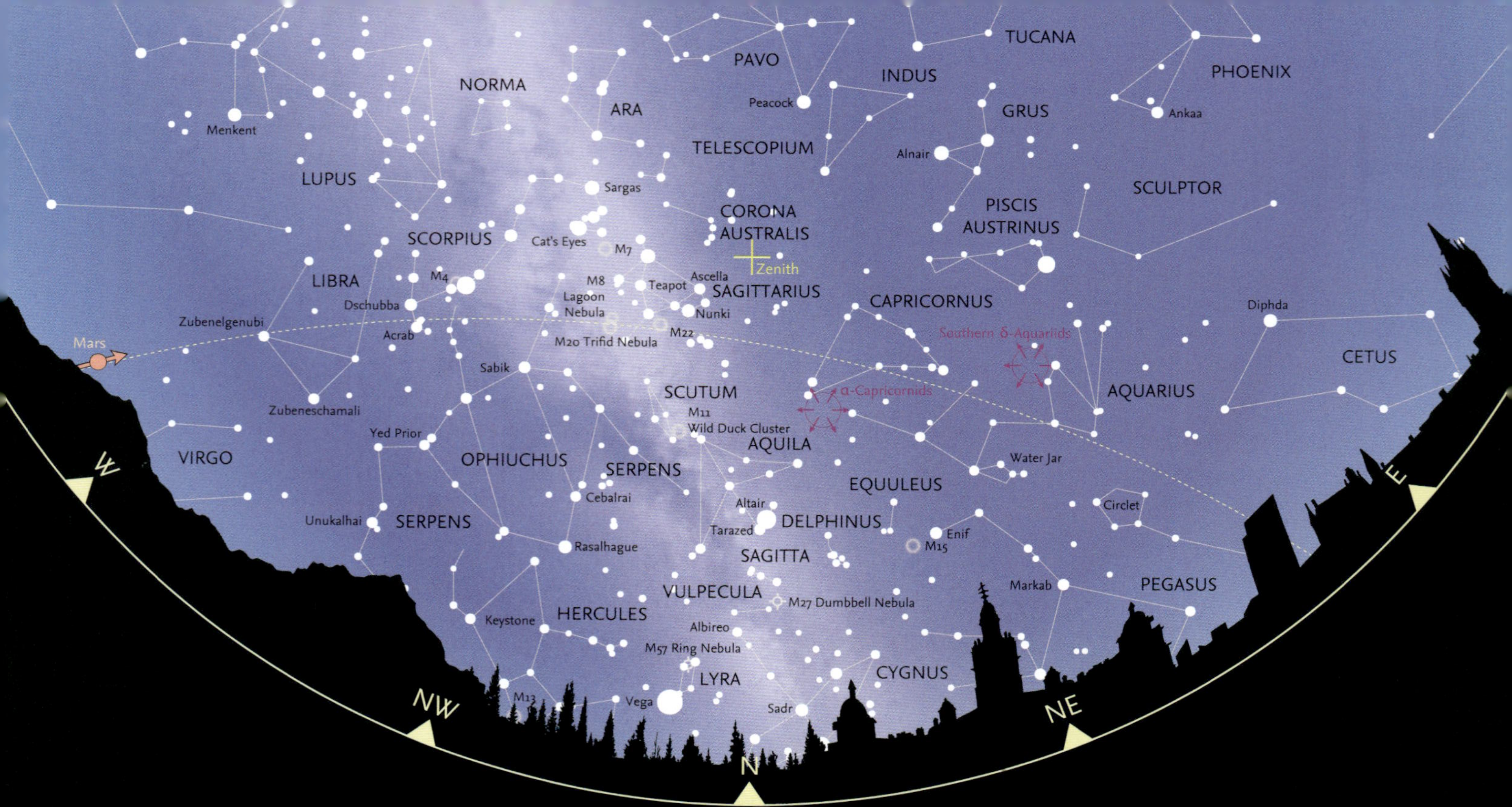

August – Looking North

1 August	23:00
15 August	22:00
1 September	21:00

August – Looking North

Cygnus, with brilliant ***Deneb*** (α Cygni), is now prominent just to the east of the meridian. The star ***Albireo*** (β Cygni) marks the Swan's beak and through a telescope it can be seen to be an optical double: the brighter star is an orange giant 100 times more luminous than the Sun; the second star is blue-white and 230 times more luminous than the Sun. The stars are not thought to be orbiting each other. The other two stars of the (northern) Summer Triangle, ***Vega*** in ***Lyra*** and ***Altair*** in ***Aquila*** are also unmistakable in the sky. West of Aquila, mainly in the star clouds of the Milky Way, lies ***Scutum***, most famous for the bright open cluster, **M11** or the Wild Duck Cluster, readily visible in binoculars. ***Hercules***, however, is beginning to descend towards the northwestern horizon. Above it, the sprawling constellation of ***Ophiuchus*** is readily seen. The Great Square of ***Pegasus*** is now visible in the east, although ***Alpheratz*** (α Andromedae) is low on the horizon. The western side of ***Pisces***, as well as ***Aquarius*** and ***Capricornus***, is fully visible along the ecliptic. ***Sagittarius*** contains many gaseous nebulae, like ***M8*** (the star-forming Lagoon Nebula, close to γ Sagittarii) and ***M20*** (the Trifid Nebula) and several globular clusters such as ***M22***. The constellation is at the zenith, with ***Scorpius*** to the west.

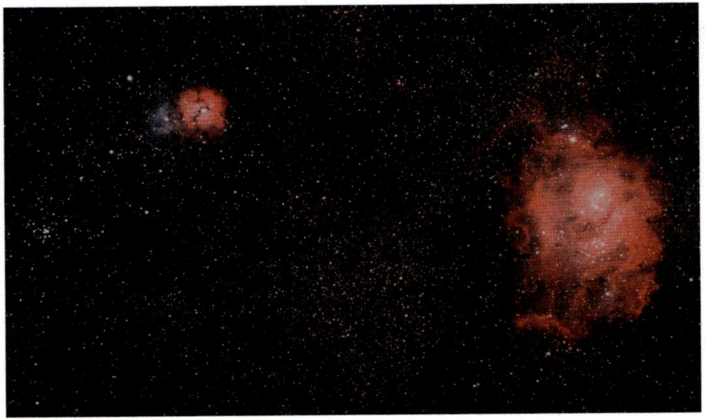

The Lagoon Nebula, a hot, glowing star-forming gas cloud.

The Moon's phases for August 2027

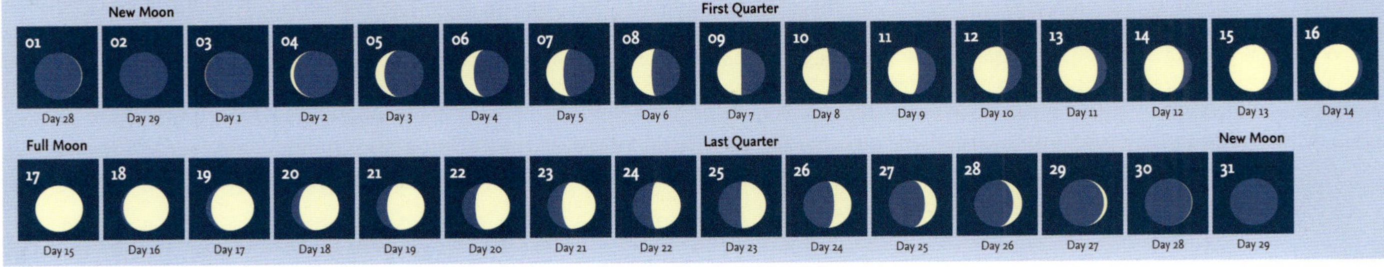

August – Moon and Planets

The Moon

The New Moon arises on 2 August, the thin crescent passes 2.4°S of ***Regulus*** and 3.0°S of dazzling ***Jupiter*** (mag. -1.7) on 3 August. On 7 August ***Spica*** can be seen 3.3°N of the waxing crescent Moon. Two days later the Moon is at First Quarter. On 11 August beautiful red-orange ***Antares*** lies only 0.1°N of the waxing gibbous Moon, visible just after sunset. The Full Moon outshines the sky on 17 August. The waning gibbous Moon passes 0.9°N of the ***Pleiades*** on 24 August, the following day it reaches Last Quarter and the month ends with a New Moon on 31 August.

The Planets

Mercury rises with the Sun until 11 August after which it swings into the evening sky, appearing just after sunset. It brightens from -1.3 to -2.0 and fades to -0.4 at the end of August. ***Venus*** is visible at dawn, rising with the Sun on 12 August after which it trails the Sun and can be seen in the evening sky, still spectacular at mag. -3.9. ***Mars*** is placed in ***Virgo***, visible in the early evening at mag. 1.3 all month. Mars passes 1.7°N of ***Spica*** on 23 August. ***Jupiter*** continues in ***Leo***, setting with the Sun towards the end of the month, at mag. -1.7. Jupiter will appear 3.0°N of the waxing crescent Moon on 3 August, with ***Regulus*** close by. ***Saturn*** lies in ***Pisces***, visible after midnight, starting the month at mag. 0.5 and slowly brightening to mag. 0.3. It enters retrograde motion on 9 August. ***Uranus*** stays in ***Taurus***, appearing shortly after midnight, starting at mag. 5.8 and brightening gradually to 5.7. ***Neptune*** is in Pisces rising an hour or so before midnight, at mag. 7.7.

The path of the Sun and the planets along the ecliptic in August.

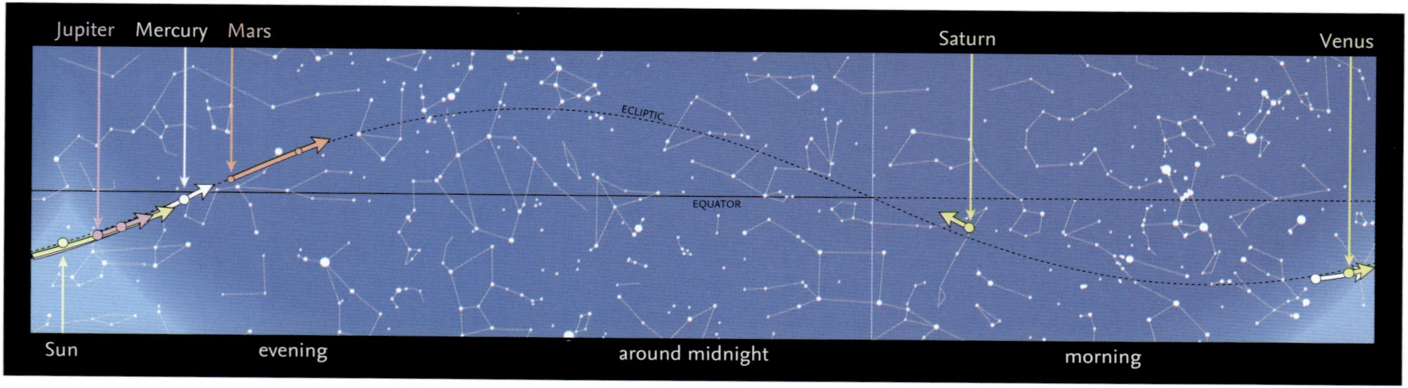

Calendar for August

02	06:28	Moon at perigee = 357,362 km
02	10:05	New Moon
02	10:06	Total solar eclipse (northern Africa, southern Saudi Arabia, Egypt). Partial eclipse visible from UK, Portugal, Spain, Madeira, Gibraltar, Sudan.
03	19:00	Regulus 2.4°N of the Moon
03	22:38	Jupiter (mag. -1.7) 3.0°N of the Moon
07	14:03	Spica 3.3°N of the Moon
09	04:54	First Quarter Moon
09	18:55	Saturn enters retrograde motion
11	01:31	Antares 0.1°N of the Moon; an occultation will be visible from Argentina, Brazil, Bolivia, Chile, Peru, Paraguay, Uruguay
12–13		Perseid meteor shower maximum
15	14:23	Moon at apogee = 406,086 km
17	07:13	Penumbral lunar eclipse; visible from the Americas, Alaska and Antarctica
17	07:29	Full Moon
23	14:36	Mars (mag. 1.3) 1.7°N of Spica
24	23:36	Pleiades 0.9°S of the Moon
25	02:27	Last Quarter Moon
30	15:36	Moon at perigee = 359,208 km

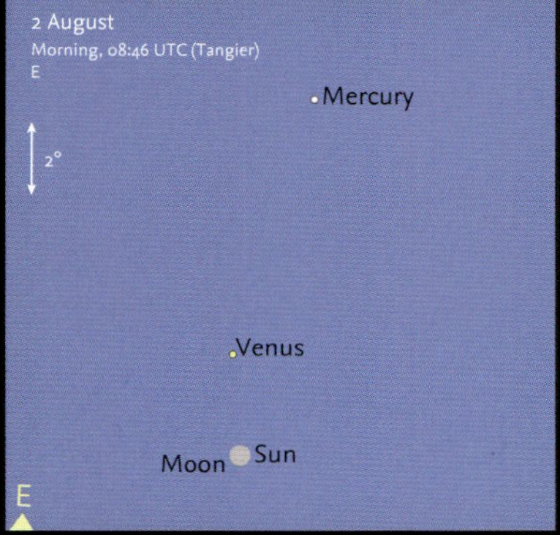

2 August • *Total solar eclipse visible from Tangier, Morocco at 08:46 UT.*

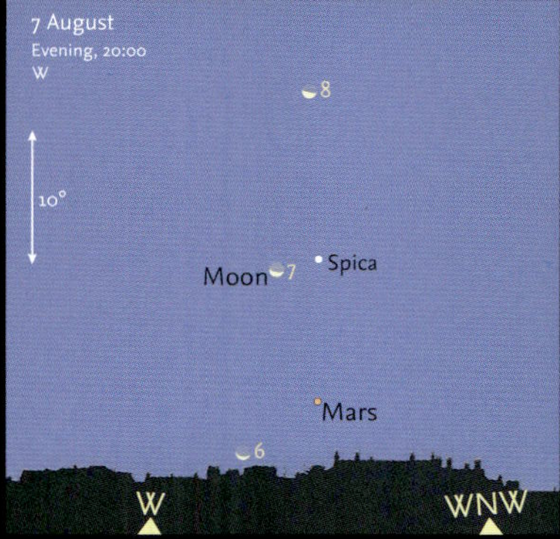

7 August • *Spica and the waxing crescent Moon, Mars nearby.*

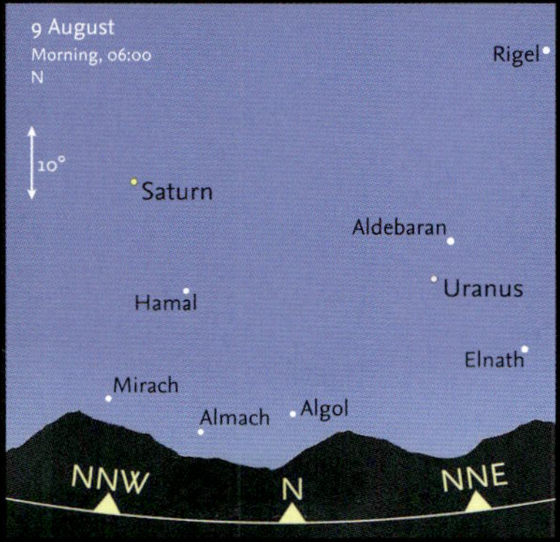

9 August • *Saturn enters retrograde motion, with Uranus close by.*

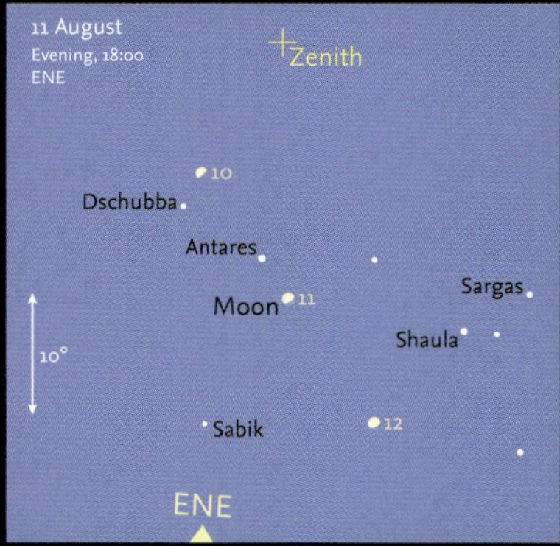

11 August • *The Moon and Antares passing close to each other.*

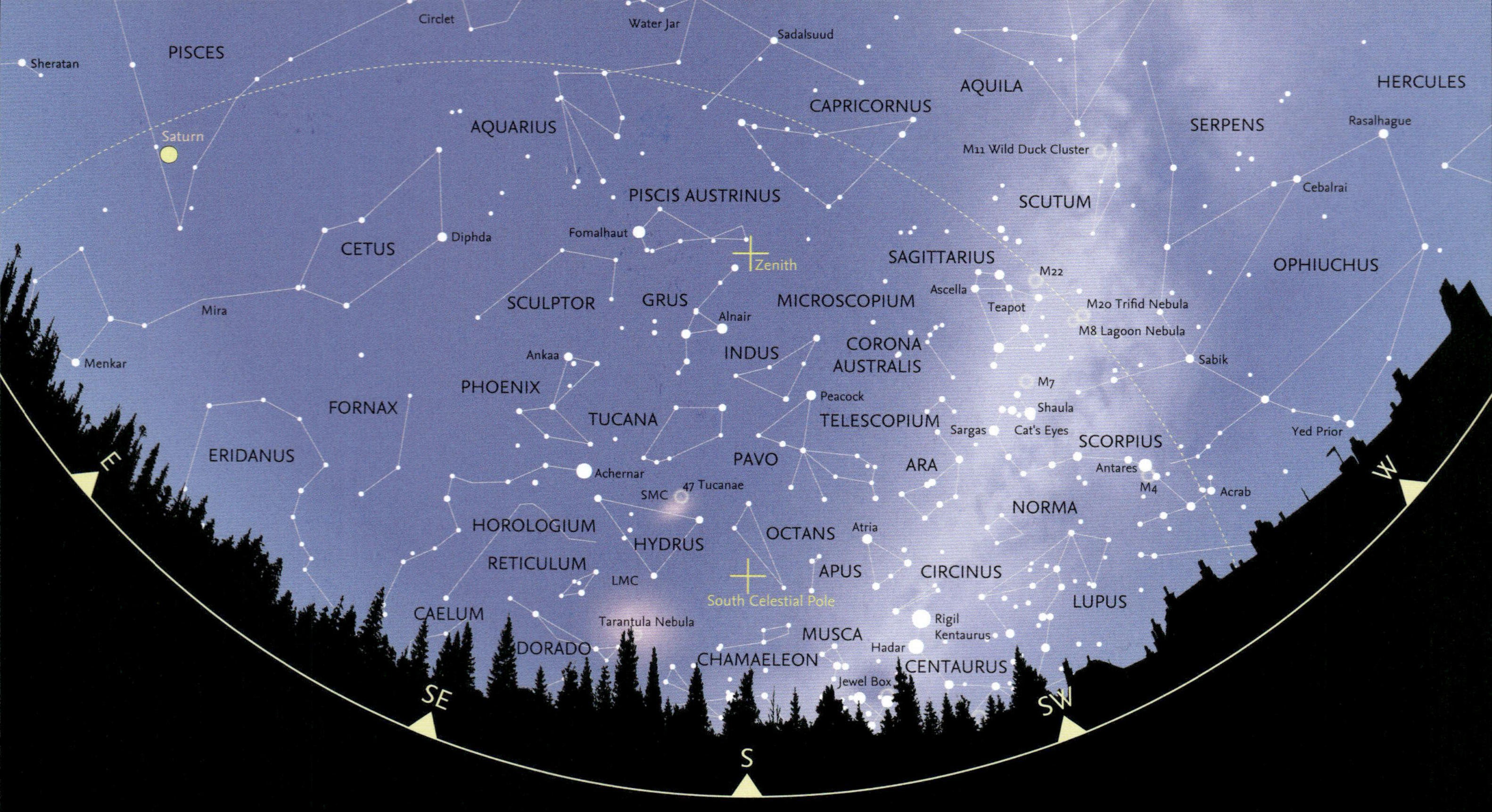

September – Looking South

1 September	23:00
15 September	22:00
1 October	21:00 (22:00 DST)

September – Looking South

The southern vernal (spring) equinox occurs on 23 September, when the Sun moves south of the equator in ***Virgo***.

The smallest of the constellations, ***Crux***, is now very low in the western sky. ***Dorado*** and the ***Large Magellanic Cloud*** (LMC) are higher and easier to see, as are the faint constellations of ***Pictor***, ***Caelum*** and ***Reticulum***. The ***Small Magellanic Cloud*** (SMC) and ***47 Tucanae*** (the globular cluster in ***Tucana***) are now nearly halfway between the horizon and the zenith. Although becoming low, the whole of ***Centaurus*** and ***Lupus*** remain visible in the southwest. ***Scorpius***, with red-orange ***Antares***, is beginning to descend in the west, but ***Sagittarius*** is clearly seen high in the sky. The constellation of ***Piscis Austrinus***, with ***Fomalhaut*** (α Piscis Austrini), is at the zenith. Below it are the constellations of ***Grus*** and ***Pavo***, with ***Achernar*** (α Eridani) and almost the whole of the long river constellation of ***Eridanus***.

Meteors

There are two minor showers in September. A few meteors may be seen from the ***α-Aurigid*** shower, active from late August. The principal peak occurs on 1 September, a day after the New Moon so viewing conditions will be favourable. At maximum, however, the rate only reaches around five meteors per hour, although the meteors are bright and relatively easy to photograph. The ***Southern Taurid*** shower begins this month (on 10 September) and, although rates are low, often produces very bright fireballs. This is a very long shower, lasting until about 20 November.

The constellation Piscis Austrinus and its brightest star, Fomalhaut.

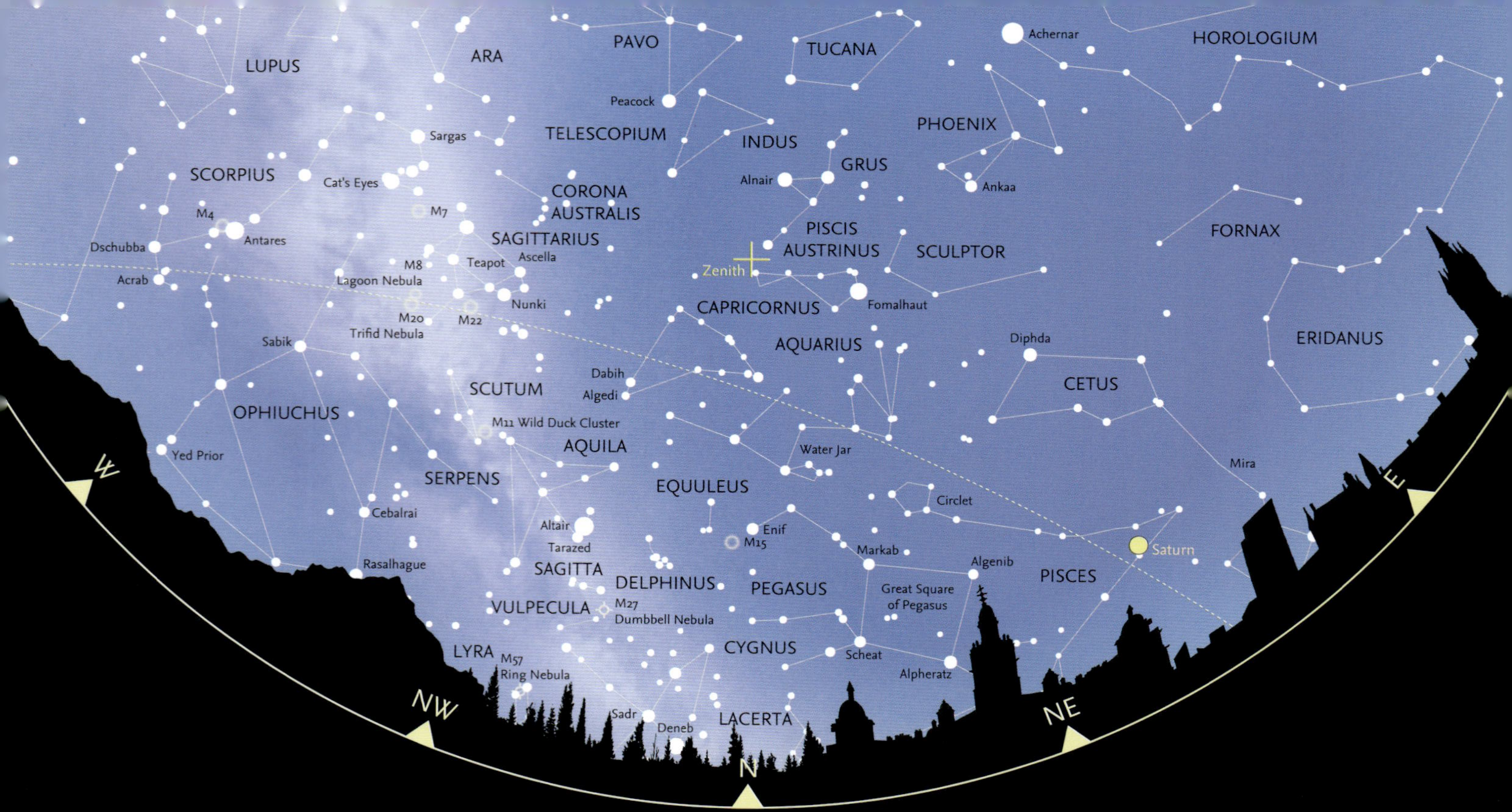

September – Looking North

1 September	23:00	
15 September	22:00	
1 October	21:00	(22:00 DST)

September – Looking North

The Great Square of ***Pegasus*** is nearing the meridian from the east and, on the west, the three bright stars of the (northern) Summer Triangle, ***Deneb*** (α Cygni), ***Altair*** (α Aquilae) and ***Vega*** (α Lyrae) are still clearly seen, although Vega is low on the horizon. ***Delphinus*** is well placed, as are the fainter constellations of ***Sagitta*** and ***Scutum*** along the Milky Way. Giant ***Ophiuchus*** is beginning to descend towards the western horizon and much of ***Libra*** has disappeared. The three zodiacal constellations of ***Pisces***, ***Aquarius*** and ***Capricornus*** are readily visible. ***Algedi*** (α Capricorni) is actually an optical double with the two stars (α^1 Cap and α^2 Cap) readily seen with the naked eye. ***Dabih*** (β Capricorni), just to the south, is also a double star and the components are relatively easy to separate with binoculars. In Aquarius, there is a small asterism consisting of four stars, resembling a tiny letter 'Y', known as the Water Jar. Somewhat higher, ***Piscis Austrinus*** is at the zenith. In classical illustrations, water is shown flowing from the Water Jar towards bright ***Fomalhaut*** (α Piscis Austrini). ***Scorpius*** and ***Sagittarius*** are now on the western side of the sky.

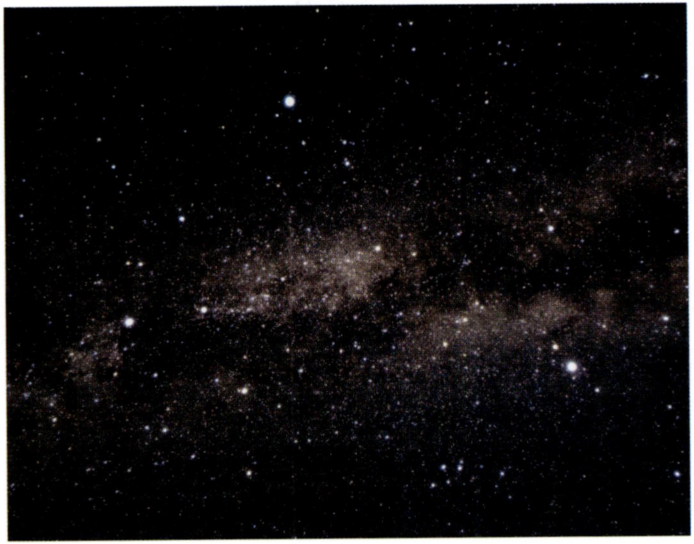

The Summer Triangle, with vertices marked by the stars Deneb, Vega and Altair.

The Moon's phases for September 2027

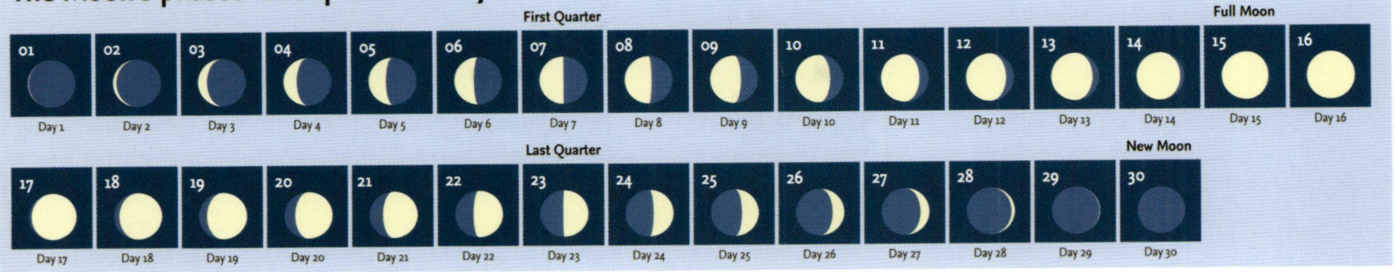

September – Moon and Planets

The Moon

Mercury (mag. -0.3) sits 4.5°N of the waxing crescent Moon on 2 September, with ***Virgo***'s brightest star ***Spica*** placed 3.2°N on 3 September, all above the horizon just after sunset. ***Antares*** lies 0.0°N of the Moon (at First Quarter) on 7 September. The Moon becomes Full on 15 September. In the early hours of 21 September the waning gibbous Moon passes 0.8°N of the ***Pleiades***, two days later it arrives at Last Quarter. On 27 September ***Regulus*** can be found 2.4°N of the waning crescent Moon, the following day ***Jupiter*** (mag. -1.7) sits 4.2°N of the Moon just before sunrise. The end of the month is marked by a New Moon.

The Planets

Mercury can be seen in the early evening just after sunset. It reaches eastern elongation on 24 September. It dims over the course of the month from -0.4 to 0.2. On 2 September Mercury (mag. -0.3) appears 4.5°N of the waxing crescent Moon. On 21 September it lies 0.2°S of ***Spica***, at mag. 0.1. ***Venus*** dominates the early evening sky at mag. -3.9. ***Mars*** progresses from ***Virgo*** to ***Libra*** after 15 September, visible in the early evening at mag. 1.3, slowly fading to 1.4. ***Jupiter*** stays in ***Leo***, setting with the Sun, at mag. -1.7 all month. On 28 September Jupiter passes 4.2°N of the waning crescent Moon. ***Saturn*** lies in ***Pisces***, climbing above the horizon late in the evening. It starts the month at mag. 0.3 and slowly brightens to mag. 0.2. ***Uranus*** stays in ***Taurus***, visible from midnight at mag. 5.7 and brightening gradually to 5.6. The ice giant enters retrograde motion on 15 September. ***Neptune*** is in Pisces visible from late evening, at mag. 7.7. Our farthest planet reaches opposition on 28 September.

The path of the Sun and the planets along the ecliptic in September.

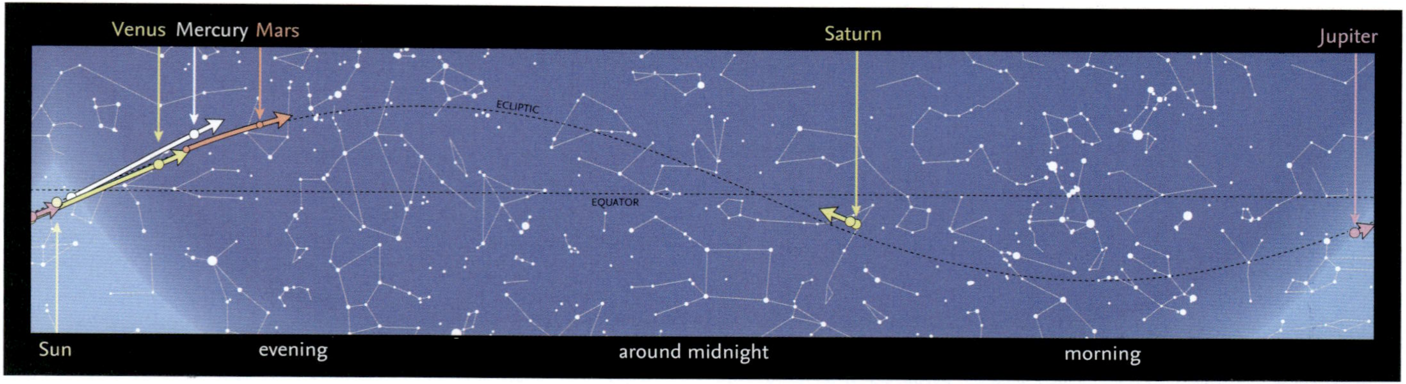

Calendar for September

01		Alpha (α) Aurigid meteor shower maximum
02	04:46	Mercury (mag. -0.3) 4.5°N of the Moon
03	23:07	Spica 3.2°N of the Moon
07	08:34	Antares 0.0°N of the Moon; an occultation will be visible from Australia, Tasmania, Papua New Guinea, Fiji, Samoa, Tonga
07	18:31	First Quarter Moon
11	23:39	Moon at apogee = 405,384 km
15	11:16	Uranus enters retrograde motion
15	23:04	Full Moon
21	03:21	Mercury (mag. 0.1) 0.2°S of Spica
21	05:12	Pleiades 0.8°S of the Moon
23	06:02	Autumnal Equinox
23	10:20	Last Quarter Moon
24	16:59	Mercury at greatest elongation (26.3°E, mag. 0.1)
27	15:35	Regulus 2.4°N of the Moon
27	20:12	Moon at perigee = 363,455 km
28	15:10	Neptune (mag. 7.7) at opposition
28	15:39	Jupiter (mag. -1.7) 4.2°N of the Moon
30	02:36	New Moon

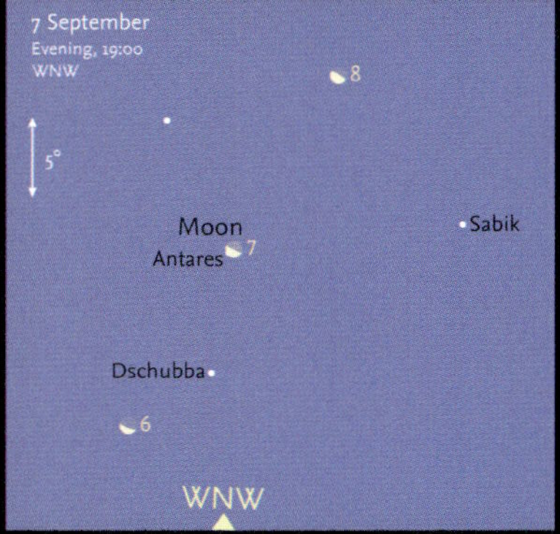

7 September • *Antares and the First Quarter Moon together in the southwest.*

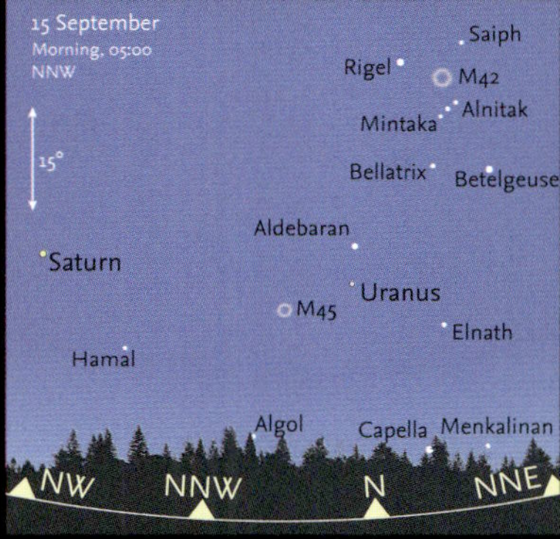

15 September • *Uranus enters retrograde motion in Taurus, the Pleiades (M45) and Aldebaran nearby.*

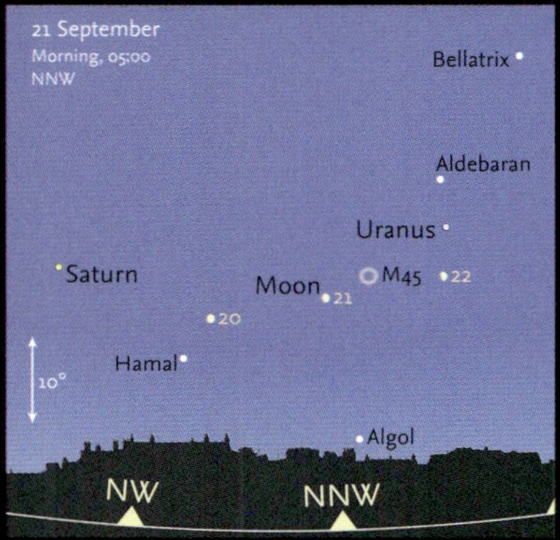

21 September • *The Pleiades and the waning gibbous Moon.*

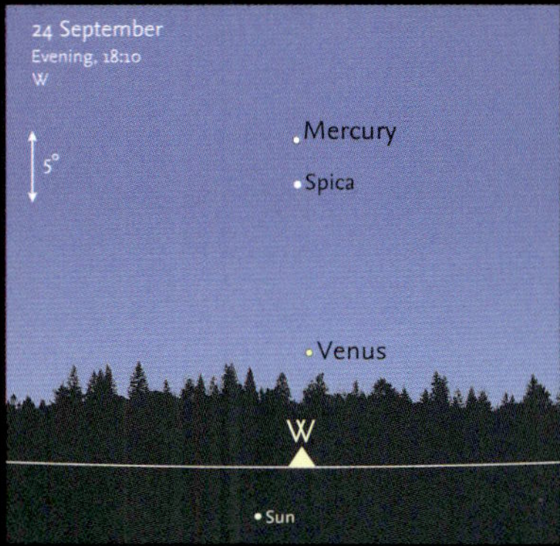

24 September • *Mercury at greatest eastern elongation with Venus to the west.*

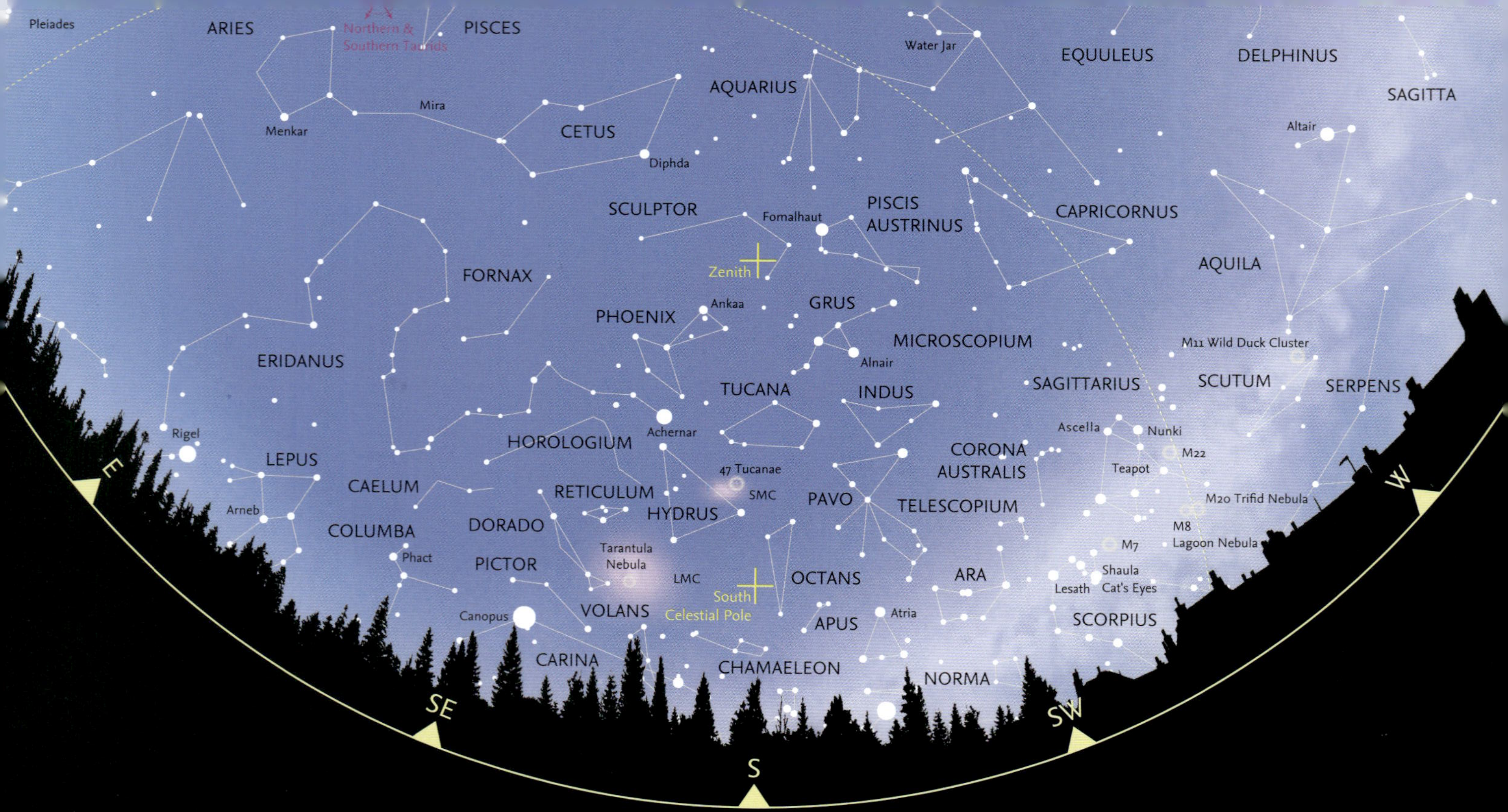

October – Looking South

1 October	23:00	(00:00 DST)
15 October	22:00	(23:00 DST)
1 November	21:00	(22:00 DST)

October – Looking South

Daylight Saving Time starts (where used) in New Zealand on Sunday 26 September and in Australia on Sunday 3 October. The ***False Cross*** has reappeared, also very low, farther towards the east on the opposite side of the meridian. ***Canopus*** (α Carinae) is now much higher, as is the ***Large Magellanic Cloud*** (LMC). The ***Small Magellanic Cloud*** (SMC) and the globular cluster ***47 Tucanae*** are on the southern meridian, halfway to the zenith, as is ***Achernar*** (α Eridani). The whole of the long constellation of ***Eridanus*** is now visible together with ***Rigel*** in ***Orion***. Still higher are ***Phoenix***, ***Grus*** and ***Piscis Austrinus***. ***Pavo*** has begun to descend in the southwest. ***Ophiuchus*** has now slipped below the horizon as has much of ***Scorpius***, only the 'tail' of which remains visible. ***Sagittarius*** is getting lower, but remains visible, as do the zodiacal constellations of ***Capricornus*** and ***Aquarius***.

Meteors

The ***Orionids*** meteor shower starts on 2 October and builds up to a broad maximum lasting around a week centred on 21–22 October. Like the May ***η-Aquariid*** shower, the Orionids are associated with Comet 1P/ Halley. During this second pass through the stream of particles from the comet, slightly fewer meteors are seen than in May. In both showers the meteors are very fast and many leave persistent trains, with hourly rates usually around 10–30, occasionally more. In 2027, there is a Last Quarter Moon lighting up the sky and interfering with observations after midnight.

The faint shower of the ***Southern Taurids*** (often with bright fireballs) continues through October, eventually reaching a peak on 5 November. Towards the end of the month (around 20 October), another shower (the ***Northern Taurids***) begins to show activity, peaking in November. The parent comet for both Taurid showers is Comet 2P/Encke. The meteors in both Taurid streams are relatively slow and bright.

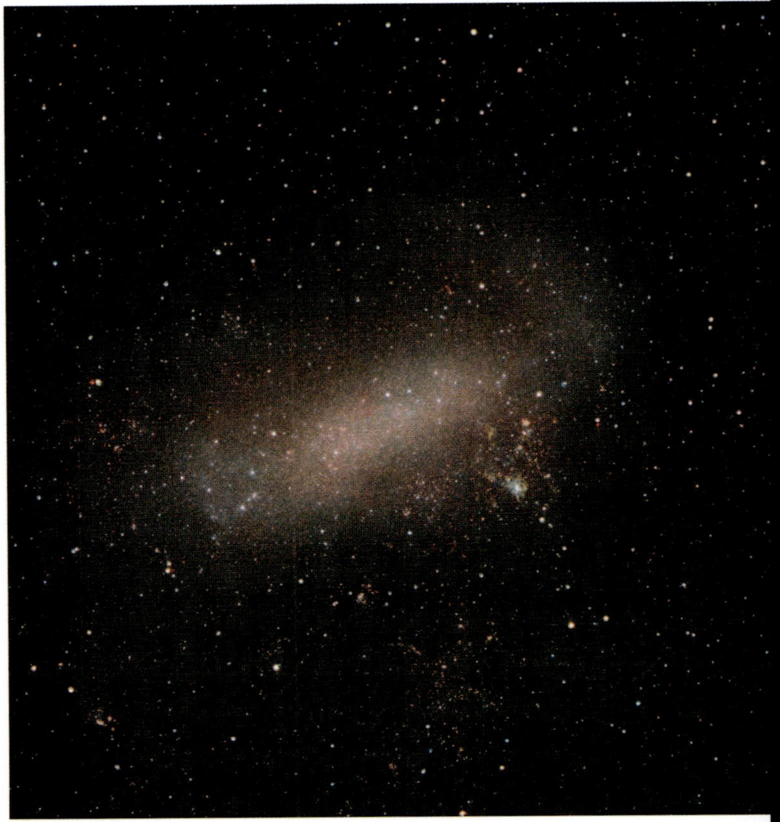

The Large Magellanic Cloud, a satellite dwarf galaxy of the Milly Way.

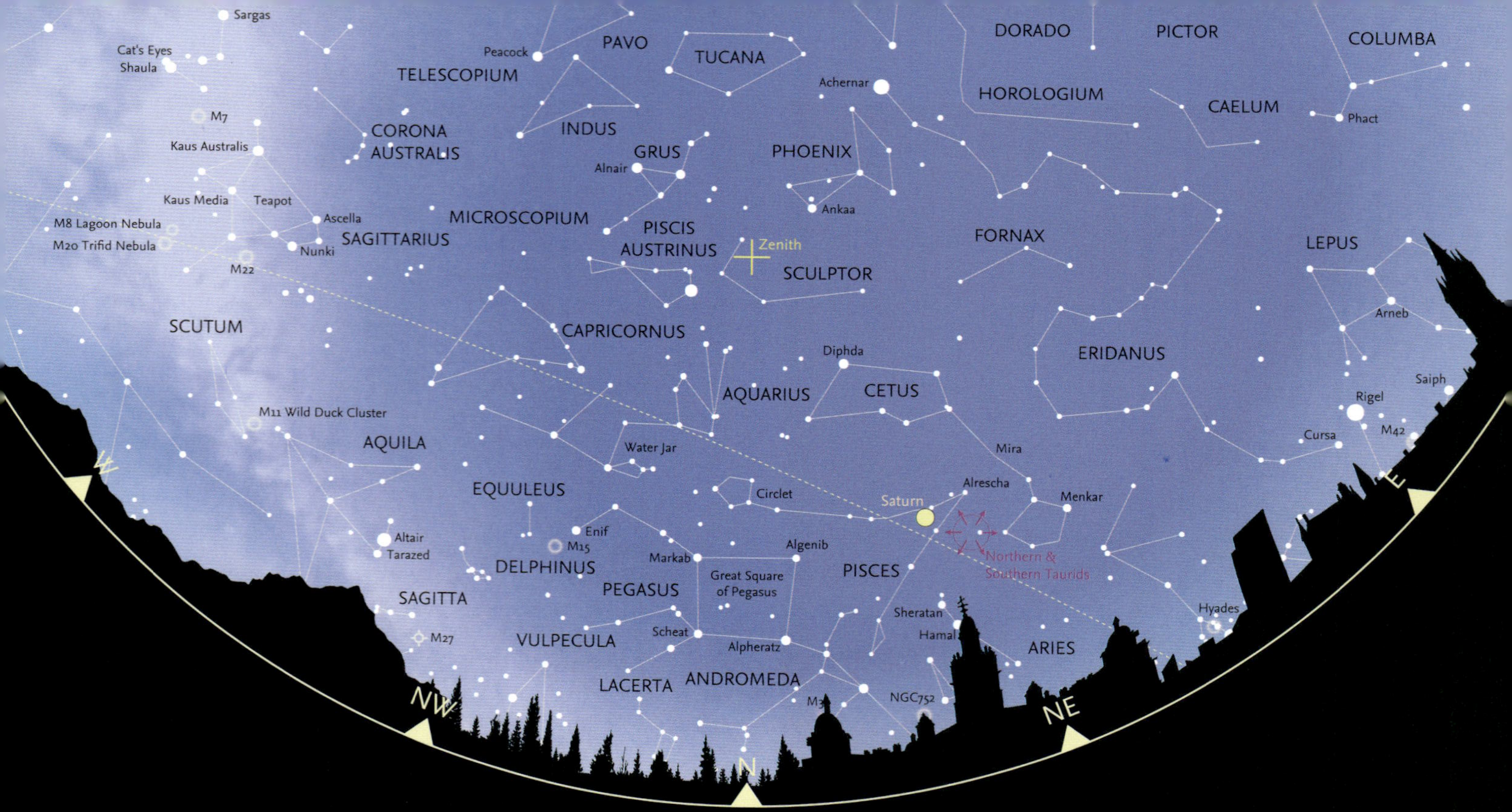

October – Looking North

1 October	23:00	(00:00 DST)
15 October	22:00	(23:00 DST)
1 November	21:00	(22:00 DST)

October – Looking North

The Great Square of ***Pegasus*** now dominates the sky to the north and the whole of ***Andromeda*** with the ***Andromeda Galaxy*** (M31) is now clear of the horizon. The two chains of stars forming ***Pisces*** frame the Great Square together with ***Alrescha*** (α Piscium) at the point where the two lines of stars join. Farther along the ecliptic, the constellations of ***Aquarius*** and ***Capricornus*** are fully visible. Farther east, all of ***Cetus*** is visible and, low down on the horizon, ***Aldebaran*** in ***Taurus*** is beginning to rise. Due east, ***Orion*** is becoming visible, with blue-white ***Rigel*** and the beginning of the long, winding constellation of ***Eridanus*** that winds its way to ***Achernar*** (α Eridani), far to the south. West of the meridian, ***Lyra***, with ***Vega*** has disappeared; much of ***Cygnus*** is invisible and ***Deneb*** is brushing the horizon. Only ***Aquila*** and ***Altair*** (α Aquilae) remain clearly visible. ***Vulpecula***, with the planetary nebula ***M27*** is still visible – as is ***Sagitta*** – and above them, the distinctive constellation of ***Delphinus*** and the very faint constellation of ***Equuleus*** (the Little Horse).

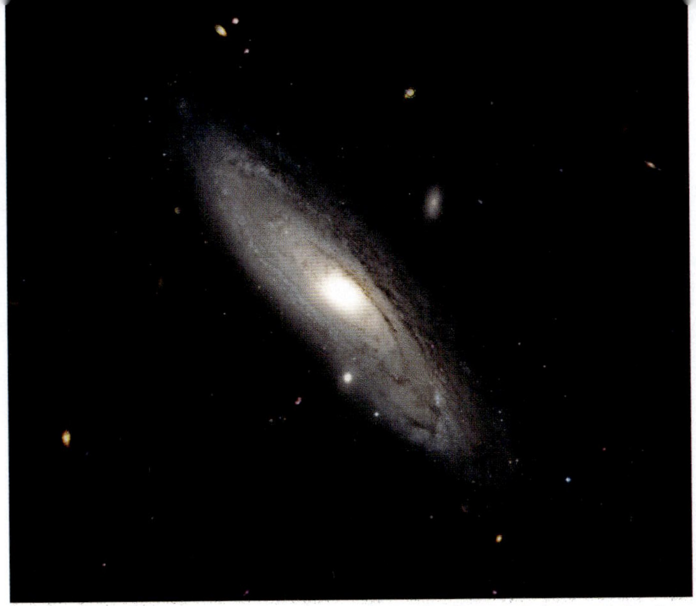

Andromeda, the nearest large spiral galaxy.

The Moon's phases for October 2027

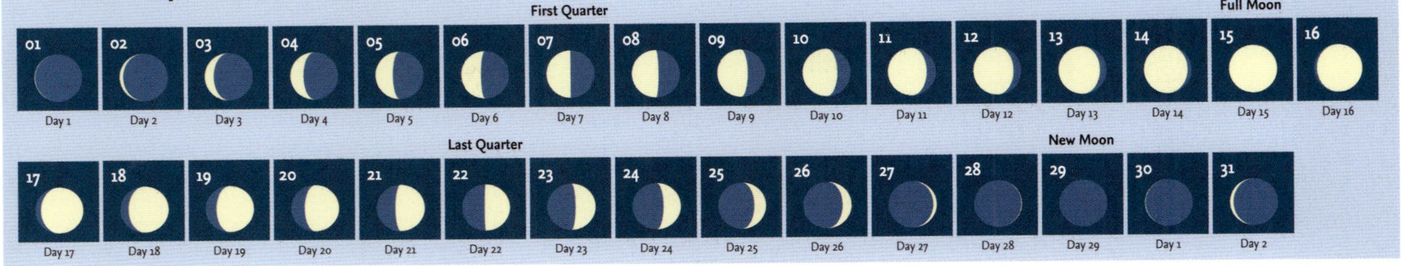

October – Moon and Planets

The Moon

The waxing crescent Moon lies 3.1°S of ***Spica*** on the first day of October, both lying very low on the horizon after sunset. On 2 October ***Mercury*** (mag. 0.3) sits 1.8°N of the Moon and the following day ***Mars*** (mag. 1.4) lies 4.5°N of the Moon, both planets visible only at dusk. On 4 October ***Antares*** will be 0.2°S of the Moon in the early evening sky. On 7 October the Moon arrives at First Quarter, on 15 October it becomes Full. On 18 October the waning gibbous Moon passes 0.6°N of the ***Pleiades.*** On 22 October a Last Quarter Moon dominates the dawn sky, two days later the waning crescent Moon passes 2.6°S of ***Regulus*** in ***Leo***, above the horizon a few hours after midnight. On 26 October ***Jupiter*** (mag. -1.8) can be found 4.9°N of the Moon before sunrise. The lunar cycle restarts on 29 October with a New Moon and the month ends with ***Venus*** (mag. -3.9) 4.4°N of the Moon, visible just after sunset.

The Planets

Mercury continues to be seen in the early evening sky, then in the early morning from 19 October. It dims dramatically at the solar conjunction and bounces back to mag. -0.1 at the end of October. Mercury at mag. 1.2 appears 3.8°S of Venus (mag. -3.9) on 11 October. ***Venus*** stays in the early evening sky at mag. -3.9. ***Mars*** moves from ***Libra*** to ***Scorpio*** from 15 October, then shifts into ***Ophiuchus*** from 27 October, low on the horizon at mag. 1.4 all month. On 29 October it appears 3.7°N of its rival red celestial object, ***Scorpio***'s brightest star – ***Antares. Jupiter*** continues in ***Leo***, visible now a few hours before sunrise, brightening from mag. -1.7 to -1.8. ***Saturn*** lies in ***Pisces***, climbing above the horizon late in the evening at mag. 0.2. It reaches opposition on 18 October, brightening to mag. 0.1 before fading back to mag. 0.2 at the end of the month. ***Uranus*** stays in ***Taurus***, above the horizon later in the evening at mag. 5.6. ***Neptune*** is in Pisces, visible from a few hours after sunset, at mag. 7.7.

The path of the Sun and the planets along the ecliptic in October.

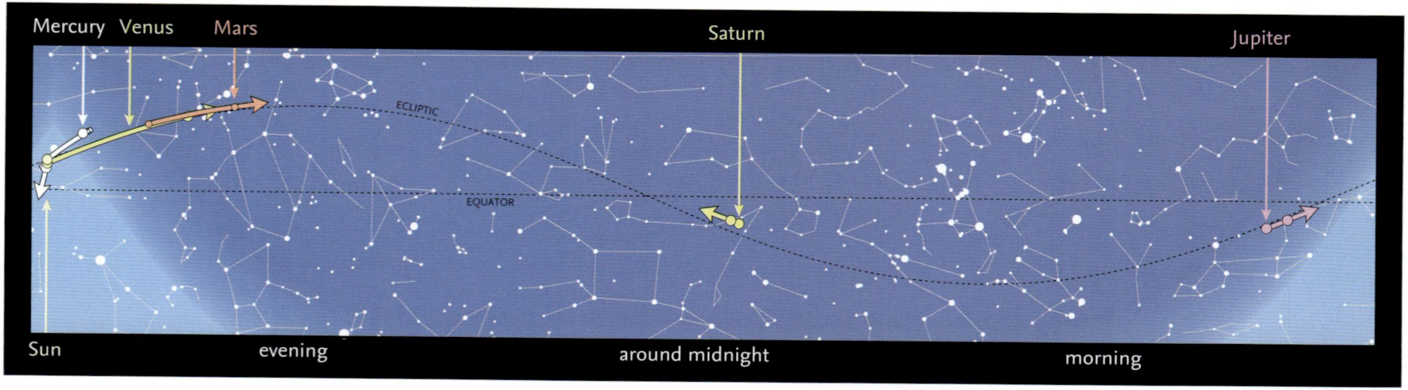

Calendar for October

01	09:12	Spica 3.1°N of the Moon
02	00:27	Mercury (mag. 0.3) 1.8°N of the Moon
03	08:58	Mars (mag. 1.4) 4.5°N of the Moon
04	17:01	Antares 0.2°S of the Moon; an occultation will be visible from Africa
07	11:47	First Quarter Moon
09	15:47	Moon at apogee = 404,531 km
11	14:00	Mercury (mag. 1.2) 3.8°S of Venus (mag. -3.9)
15	13:47	Full Moon
18	01:27	Saturn (mag. 0.1) at opposition
18	10:48	Pleiades 0.6°S of the Moon
21–22		Orionid meteor shower maximum
22	00:00	Last Quarter Moon
24	23:09	Regulus 2.6°N of the Moon
25	05:36	Moon at perigee = 368,651 km
26	09:33	Jupiter (mag. -1.8) 4.9°N of the Moon
29	13:36	New Moon
29	13:47	Mars (mag. 1.4) 3.7°N of Antares

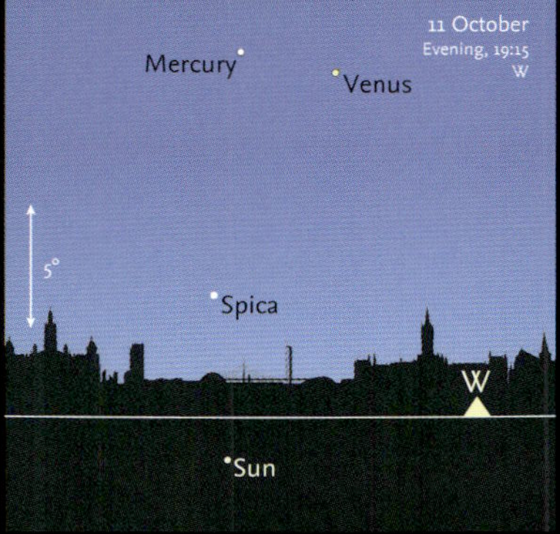

11 October • *Mercury and Venus setting together.*

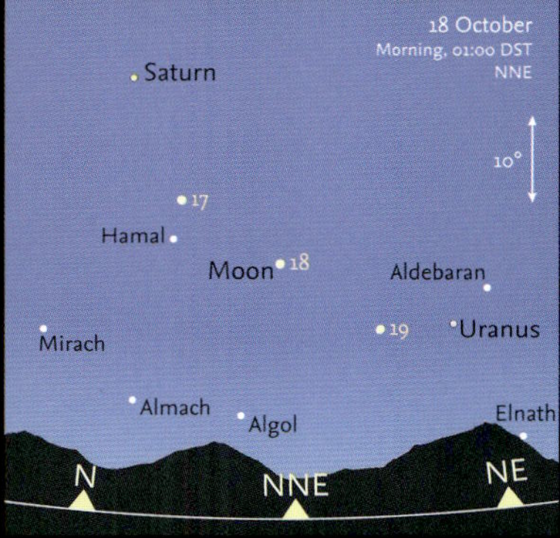

18 October • *Saturn at opposition. The waning gibbous Moon, Uranus and Aldebaran lie to the east.*

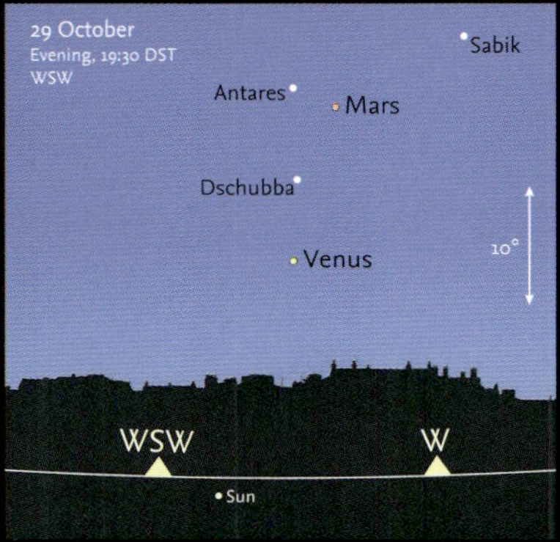

29 October • *Mars and its red rival, Antares.*

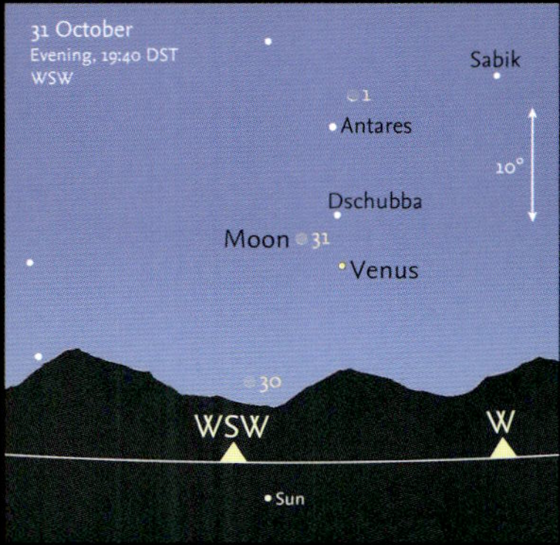

31 October • *Venus close to the thin waxing crescent Moon.*

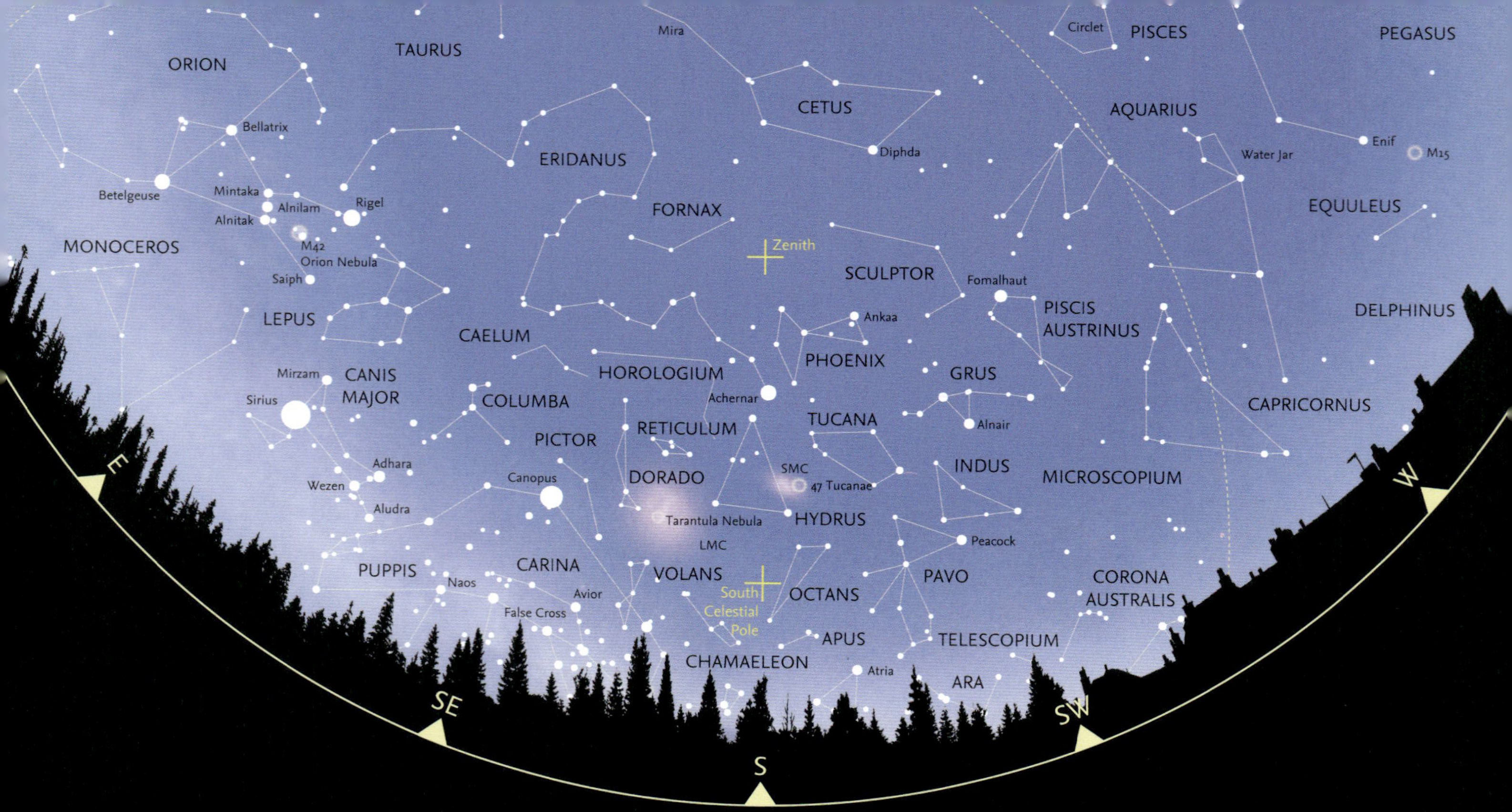

November – Looking South

1 November	23:00	(00:00 DST)
15 November	22:00	(23:00 DST)
1 December	21:00	(22:00 DST)

November – Looking South

The ***False Cross*** on the ***Carina/Vela*** border is now higher, and the constellation of ***Puppis*** as well as ***Canopus*** (α Carinae), and both the ***Large Magellanic Cloud*** (LMC) and the ***Small Magellanic Cloud*** (SMC), are clearly visible. ***Achernar*** (α Eridani) is halfway between the South Celestial Pole and the zenith. The whole of ***Eridanus***, which starts near ***Rigel*** in ***Orion***, is now clearly seen as it winds its way to Achernar. ***Pavo*** is becoming lower in the southwest and, in the west, most of ***Sagittarius*** is below the horizon, with ***Capricornus*** descending behind it. ***Corona Australis*** is still just visible. In the east, ***Canis Major*** is now clearly seen, together with the small constellations of ***Columba*** and ***Lepus*** above it.

Meteors

Two meteor showers begin in September or October but continue into November. The ***Orionids*** (see page 89), one of the streams associated with Comet 1P/Halley, continue until about 7 November. The ***Southern Taurids*** (see page 83) begin on 10 September and continue until 20 November, reaching a peak around 5 November. The ***Northern Taurid*** shower, which began in mid-October, reaches maximum – although only with a rate of about five meteors per hour – on 12–13 November. The Moon is close to First Quarter on 5 November and sets a few hours before midnight and it's even brighter in its waxing gibbous phase on 12 and 13 November and will dominate the sky throughout the night. The shower gradually trails off, ending around 10 December.

The ***Leonid*** shower is best seen from the northern hemisphere due to the position of its radiant, but southern observers may see some rising from the horizon. They have a short period of activity (6–30 November), with maximum on 18 November. This shower is associated with Comet 55P/Tempel-Tuttle and has shown extraordinary activity on various occasions with many thousands of meteors per hour. The rate is likely to be about 10–15 per hour at peak activity; however, the Moon will be waning gibbous and above the horizon all night. These meteors are the fastest shower meteors recorded (about 70 km per second) and often leave persistent trains.

There is a minor southern meteor shower that begins activity in late November. This is the ***Phoenicids***, but little is known of the shower, partly because the parent comet is believed to be the disintegrated Comet D/1819 W1 (Blanpain). With no accurate knowledge of the location of the remnants of the comet, predicting the possible rate becomes little more than guesswork, but the rate is variable and may rapidly increase (as might be expected) if the orbit is nearby. Bright meteors tend to be quite frequent and the meteors are fairly slow. The radiant is located within ***Phoenix***, not far from the border with ***Eridanus*** and the bright star ***Achernar*** (α Eridani).

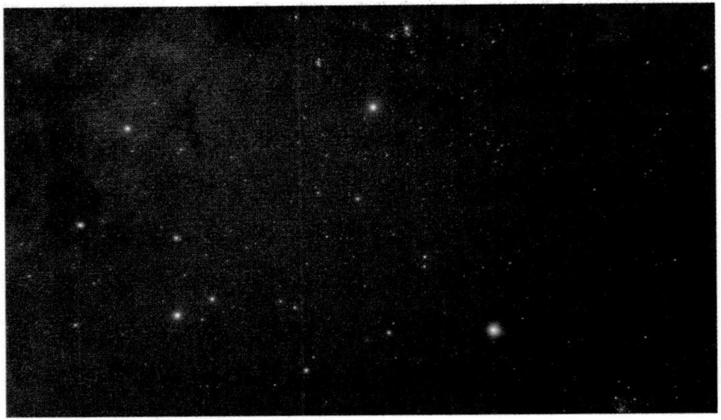

The False Cross, an asterism in the southern sky.

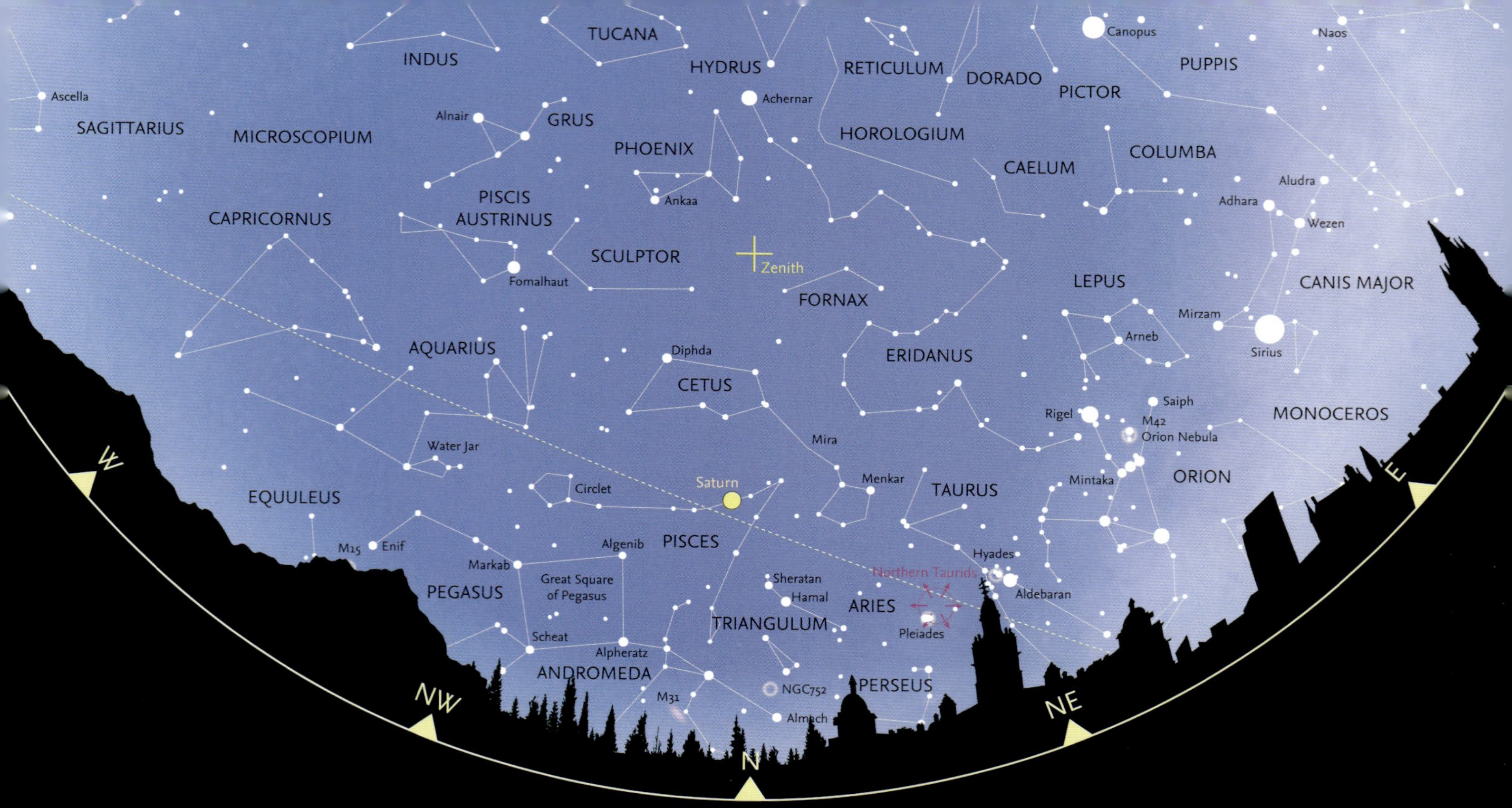

November – Looking North

1 November	23:00	(00:00 DST)
15 November	22:00	(23:00 DST)
1 December	21:00	(22:00 DST)

November – Looking North

Andromeda is now due north, and the ***Andromeda Galaxy*** (M31) can be clearly seen. The constellation of ***Triangulum*** lies between Andromeda and the zodiacal constellation of ***Aries***. Much of ***Perseus*** (including the variable star, ***Algol***) is now above the horizon, together with part of ***Auriga*** rising later in the evening. The whole of ***Pegasus***, with the Great Square, is visible and above it the two lines of stars forming ***Pisces*** and the Circlet. Higher still is the constellation of ***Cetus*** with the famous variable star, ***Mira*** (ο Ceti), with a typical range of magnitude 3.4 to 9.5. ***Taurus*** with ***Aldebaran*** (α Tauri), and the ***Pleiades*** and the ***Hyades*** clusters are visible in the southeast. ***Orion*** has fully risen in the east with ***Lepus*** above it. The long, winding constellation of ***Eridanus*** begins near ***Rigel*** in Orion. The faint constellation of ***Monoceros*** lies to the east of Orion and straddles the Milky Way. ***Canis Major*** and brilliant ***Sirius*** are even farther towards the east. In the west, ***Aquarius*** (with its distinctive asterism of the Water Jar) and ***Capricornus*** are clearly seen, with ***Piscis Austrinus*** and bright ***Fomalhaut*** (α Piscis Austrini) higher in the sky. The faint constellation of ***Sculptor*** lies between Piscis Austrinus and the zenith.

The constellation Pisces.

The Moon's phases for November 2027

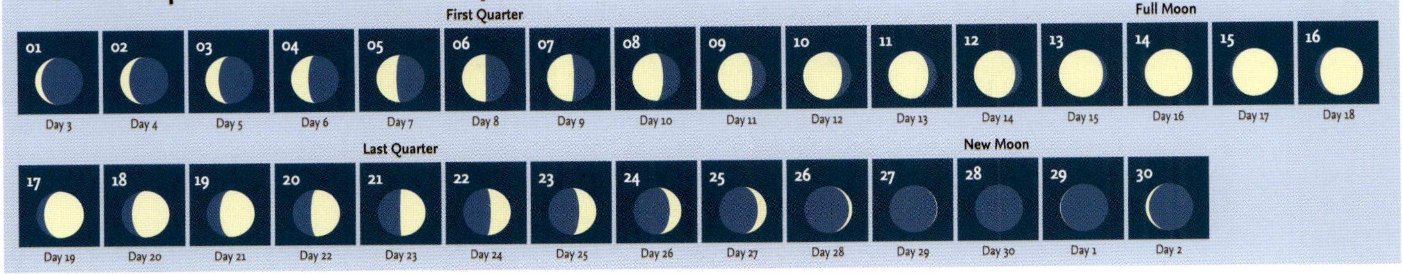

November – Moon and Planets

The Moon

On 1 November the waxing crescent Moon sits 0.4°N of orange-red ***Antares*** and 3.3°S of ***Mars*** (mag. 1.3), all three low in the western sky at sunset. On 6 November there will be a First Quarter Moon, it becomes Full on 14 November and passes 0.5°N of the ***Pleiades***. The Last Quarter Moon arrives on 21 November and passes 2.9°S of ***Regulus*** before dawn. On 25 November ***Spica*** lies 3.2°N of the Moon a few hours before sunrise, three days later the Moon is New. The end of the month is marked by two planets sitting close to the waxing crescent Moon at dusk: ***Mars*** (mag. 1.3) lies 1.6°N and ***Venus*** (mag. -3.9) 0.9°N of the Moon.

The Planets

Mercury is visible at dawn, it climbs away from the Sun to western elongation on 4 November. It brightens from mag. -0.2 to -0.9 over the course of the month. On 5 November Mercury (mag. -0.6) passes 4.0°N of ***Spica*** in ***Virgo***. ***Venus*** is spectacular in the evening sky at mag. -3.9. On 9 November Venus lies 3.8°N of ***Antares***. On 25 November it sits 0.3°S of ***Mars*** (mag. 1.3) and on 30 November Venus passes 0.9°N of the waxing crescent Moon. Mars is in ***Ophiuchus*** and moves into ***Sagittarius*** from 21 November, low in the sky just after sunset at mag. 1.3 all month. Mars will appear 3.3°N of the waxing crescent Moon on the first day of the month and it approaches the waxing crescent Moon again on the last day of November, at 1.6°N. ***Jupiter*** moves from its long stay in ***Leo*** to Virgo on 23 November, appearing above the horizon several hours after midnight, brightening from mag. -1.8 to -1.9. ***Saturn*** lies in ***Pisces***, visible from sunset, fading slightly from mag. 0.2 to 0.3. ***Uranus*** stays in ***Taurus***, above the horizon a few hours after sunset, at mag. 5.6. It reaches opposition on 30 November. ***Neptune*** is in Pisces, visible from sunset, at mag. 7.7.

The path of the Sun and the planets along the ecliptic in November.

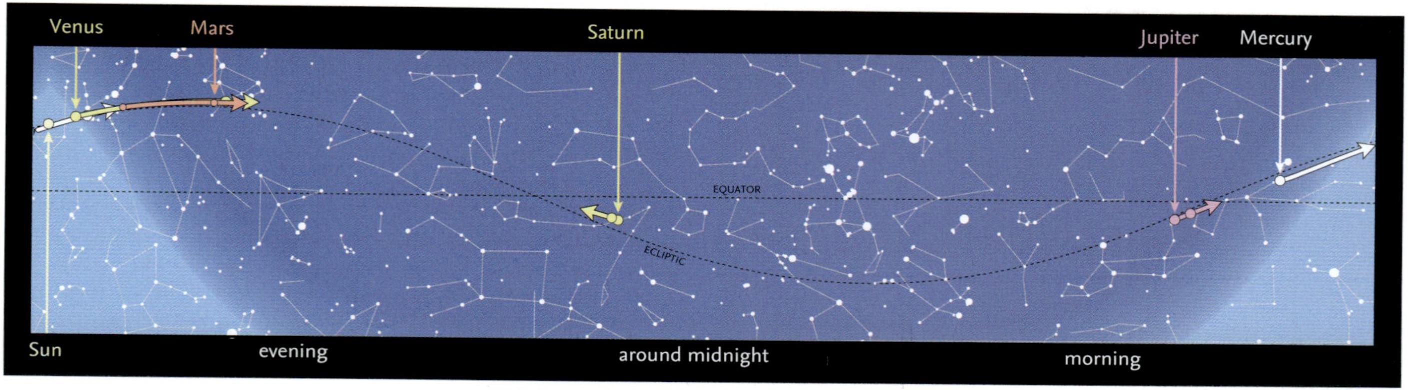

Calendar for November

01	02:13	Antares 0.4°S of the Moon
01	06:53	Mars (mag. 1.3) 3.3°N of the Moon
04	07:58	Mercury at greatest elongation (18.8°W, mag. -0.5)
05	00:48	Mercury (mag. -0.6) 4.0°N of Spica
05		Southern Taurid meteor shower maximum
06	08:00	First Quarter Moon
06	11:37	Moon at apogee = 404,183 km
09	06:44	Venus (mag. -3.9) 3.8°N of Antares
12–13		Northern Taurid meteor shower maximum
14	03:26	Full Moon
14	18:14	Pleiades 0.5°S of the Moon
17–18		Leonid meteor shower
19	00:10	Moon at perigee = 369,356 km
21	00:48	Last Quarter Moon
21	04:52	Regulus 2.9°N of the Moon
25	01:00	Venus (mag. -3.9) 0.3°S of Mars (mag. 1.3)
25	02:24	Spica 3.2°N of the Moon
28	03:24	New Moon
30	07:53	Mars (mag. 1.3) 1.6°N of the Moon
30	09:13	Uranus (mag. 5.6) at opposition

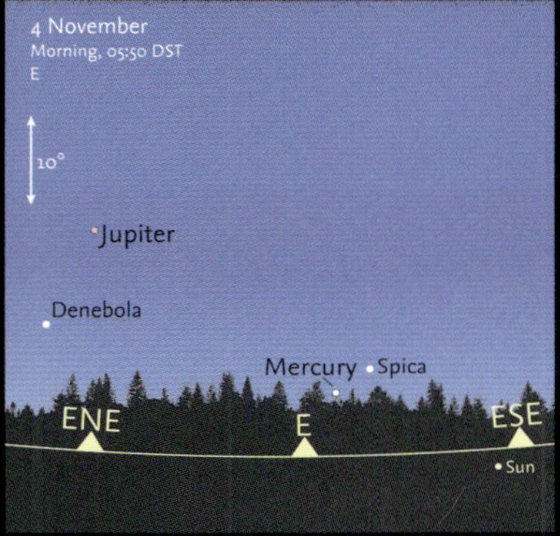

4 November • *Mercury at greatest western elongation. Jupiter sits nearby.*

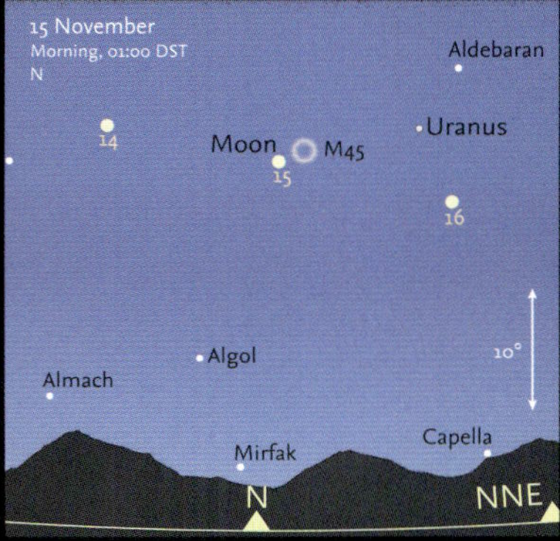

15 November • *The Pleiades (M45) and the Full Moon together, with Uranus and Aldebaran close by.*

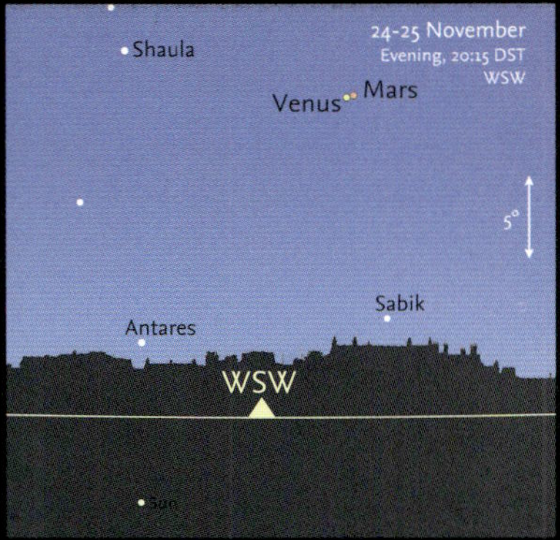

24–25 November • *Venus and Mars sit next to each other.*

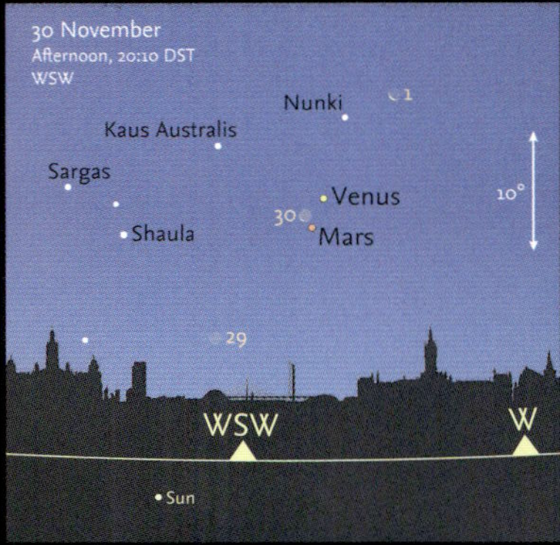

30 November • *Venus and Mars by the thin waxing crescent Moon.*

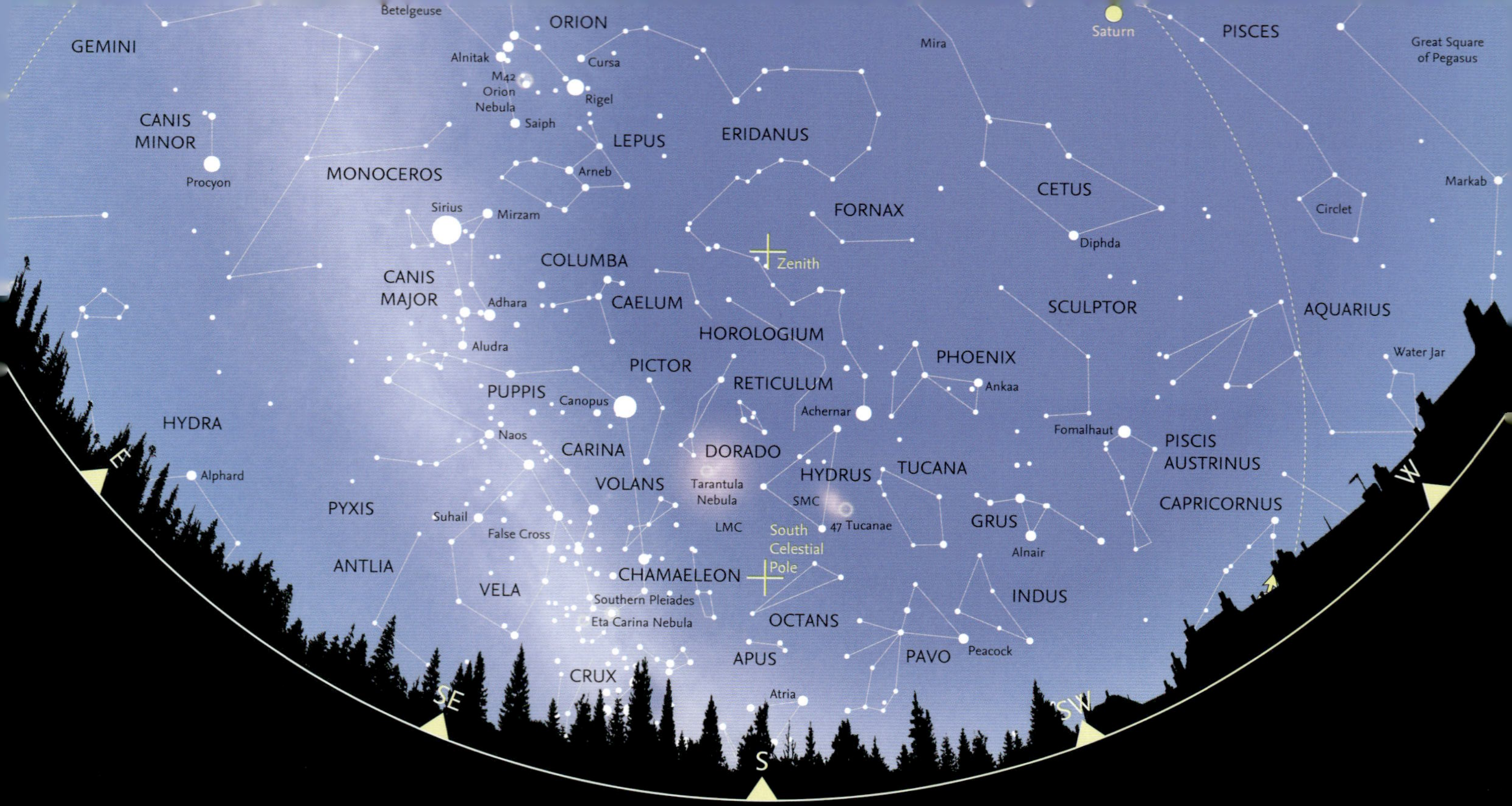

December – Looking South

1 December 23:00 (00:00 DST)
15 December 22:00 (23:00 DST)
1 January 21:00 (22:00 DST)

December – Looking South

Around summer solstice on 22 December southern hemisphere observers experience the longest period of daylight (midsummer's day). ***Crux*** and the two brightest stars in ***Centaurus***, ***Hadar*** and ***Rigil Kentaurus***, are now higher above the horizon. The ***Eta Carina Nebula*** and the ***Southern Pleiades*** are now conveniently placed for observation. Above them, the ***False Cross*** is clearly seen, with the whole of ***Vela*** and below it the constellation of ***Antlia*** (the air pump). ***Carina*** with ***Canopus*** (α Carinae) and ***Puppis*** are roughly halfway between the horizon and the zenith. ***Sirius*** and ***Canis Major*** are high in the east. In the west, ***Achernar*** (α Eridani) and ***Phoenix*** are about the same altitude as Canopus. The faint constellations of ***Pictor***, ***Dorado***, ***Reticulum*** and ***Horologium*** lie between them. ***Pavo*** with its brightest star ***Peacock*** (α Pavonis) is becoming low, as are the constellations of ***Indus***, ***Grus*** and ***Piscis Austrinus***. Higher still are the constellations of ***Sculptor*** and, near the zenith, ***Fornax***. ***Capricornus*** is largely invisible, but most of ***Aquarius*** may still be seen.

Meteors

The ***Phoenicid*** shower continues into December, reaching its weak maximum on 2 December. The ***Puppid-Velid*** shower's radiant is on the border between the two constellations. The shower begins on 1 December, lasting until 15 December, with maximum on 7 December. It is a weak shower with a maximum hourly rate of about 10 meteors, but bright meteors are often seen. There is one significant meteor shower in December (the last major shower of the year). This is the ***Geminid*** shower, which is visible over the period 4–20 December and comes to maximum on 14 December, one day after the Full Moon, so significant moonlight will affect observing conditions. It is one of the most active showers of the year, with a peak rate of around 100 meteors per hour. It is the one major shower that shows good activity before midnight. The source of the Geminids is debris from an asteroid called 3200 Phaethon; most meteor showers originate from cometary debris. The Geminids are assumed to consist of denser, rocky material, they are slower than most other meteors and often appear to last longer. The brightest often break up into numerous luminous fragments that follow similar paths across the sky.

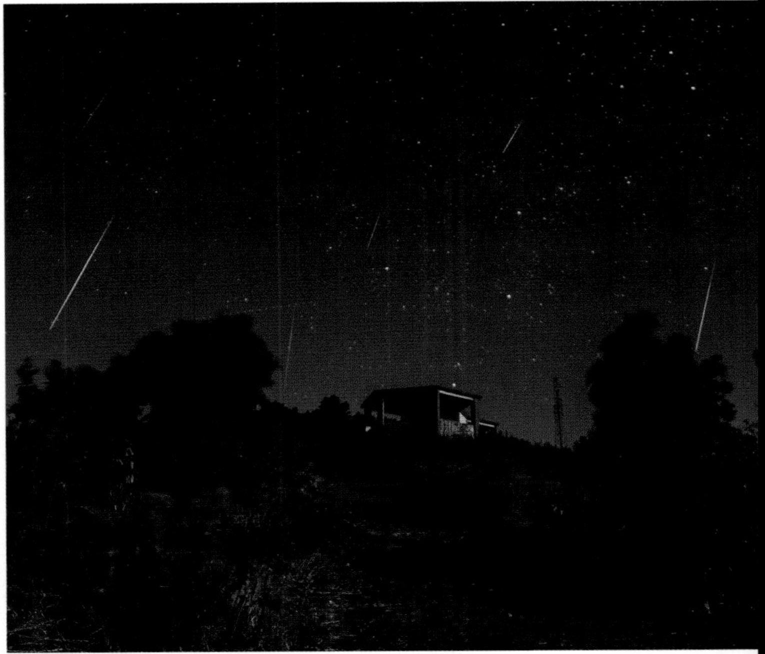

The Geminid meteor shower.

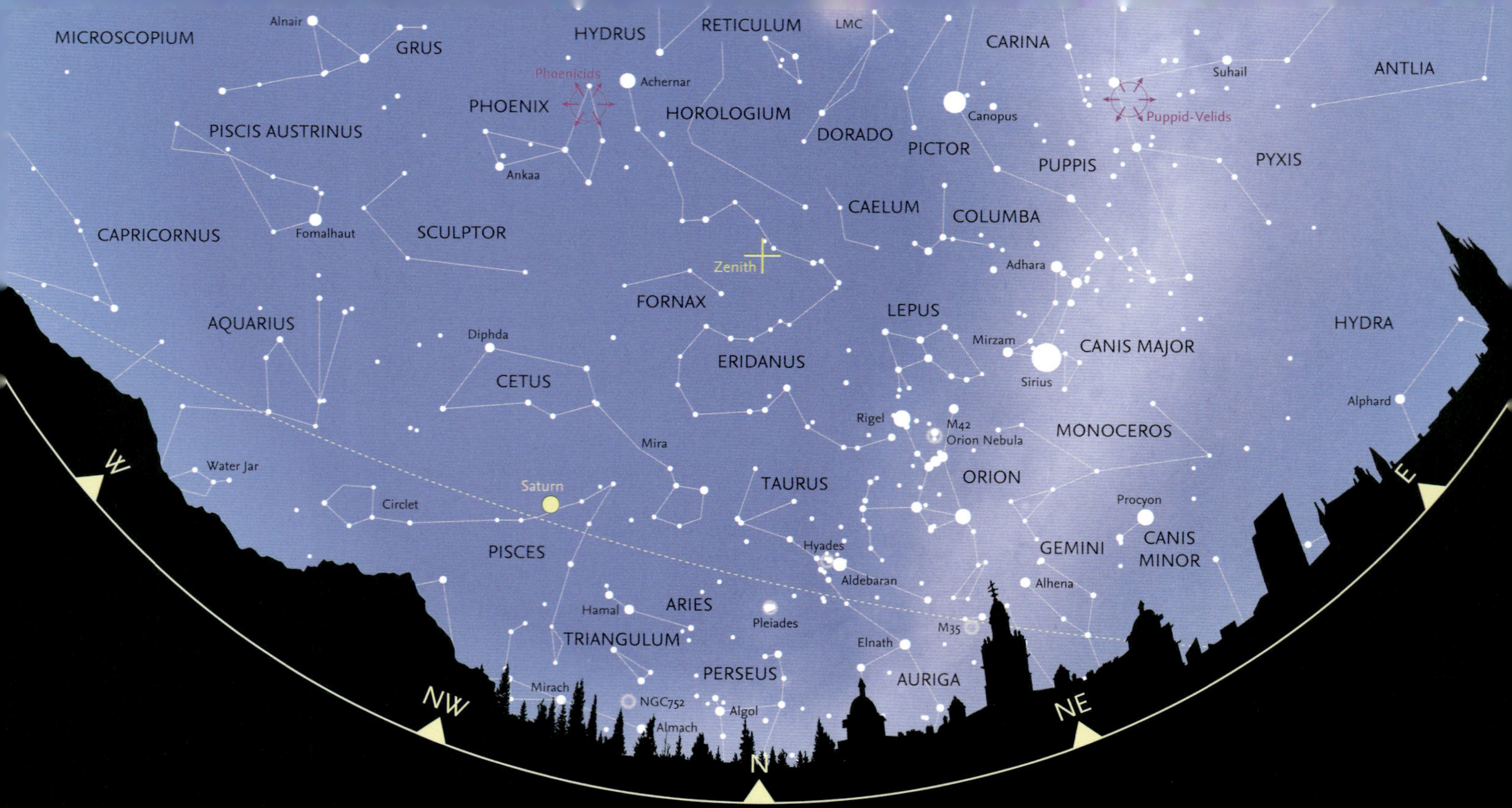

December – Looking North

1 December	23:00	(00:00 DST)
15 December	22:00	(23:00 DST)
1 January	21:00	(22:00 DST)

December – Looking North

Most of ***Perseus*** may be seen due north with, above it, the beautiful open cluster of the ***Pleiades***, the ***Hyades*** cluster, the orange star ***Aldebaran*** and the rest of ***Taurus***. Farther west, ***Andromeda*** is very low, and the southern stars of ***Pegasus*** have been lost below the horizon. Above Pegasus lies the zodiacal constellation of ***Pisces*** and, still higher, ***Cetus*** and the faint constellations of ***Sculptor*** and ***Fornax***. The famous variable of ***Mira*** in Cetus (see charts on pages 100 and 102) is ideally placed for observation. Towards the east, the whole of the constellations of both ***Auriga*** and ***Gemini*** are visible, although ***Capella*** (α Aurigae) and ***Castor*** and ***Pollux*** (α and β Gemini) are low on the horizon. ***Orion***, ***Canis Major*** (with ***Sirius***, the brightest star in the sky) and ***Canis Minor*** (with ***Procyon***) are readily visible, together with the faint constellation of ***Monoceros***. ***Eridanus*** wanders from its start near ***Rigel*** (β Orionis) towards ***Achernar*** (α Eridani), beyond the zenith.

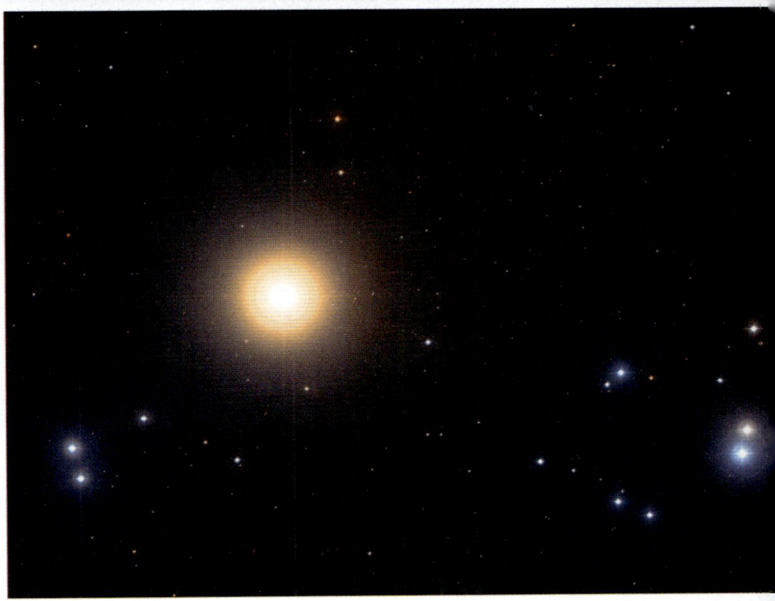

The red supergiant Betelgeuse.

The Moon's phases for December 2027

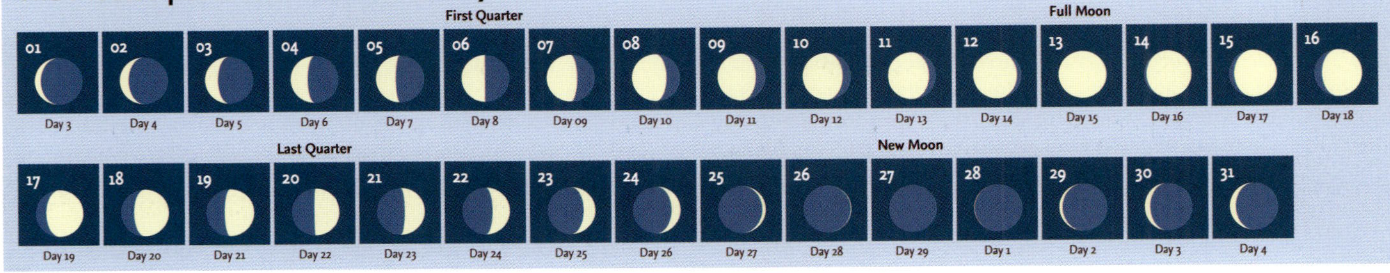

December – Moon and Planets

The Moon

The First Quarter Moon arises on 6 December. The waxing gibbous Moon passes 0.5°N of the ***Pleiades*** on 12 December a few hours before dawn. On 13 December the Moon becomes Full. ***Regulus*** lies 3.1°N of the waning gibbous Moon on 18 December, best seen a few hours before sunrise. Two days later the Moon arrives at Last Quarter. On 22 December ***Spica*** in ***Virgo*** sits 3.3°N of the waning crescent Moon before dawn. On 25 December ***Antares*** can be found 0.4°S of the Moon visible just before sunrise and two days later the lunar phase cycle resets with a New Moon. On 29 December ***Mars*** (mag. 1.2) passes 0.5°S of the Moon and on the last day of December 2027 ***Venus*** (mag. -4.0) will be found 3.1°S of the Moon. Both events can be observed at sunset.

The Planets

Mercury switches from the dawn to dusk sky on 11 December. It brightens from mag. -0.1 to -1.3 and fades to -0.9 over the course of the month. ***Venus*** is beautiful in the evening sky, brightening from mag. -3.9 to -4.0. Venus appears 3.1°S of the waxing crescent Moon on 31 December. ***Mars*** is in ***Sagittarius*** and sets with the Sun; it starts the month at mag. 1.3, brightening to 1.2. On 29 December Mars sits 0.5°S of the thin waxing crescent Moon. ***Jupiter*** is in ***Virgo*** and pops above the horizon an hour after midnight, brightening from mag. -1.9 to -2.1. ***Saturn*** lies in ***Pisces***, visible from sunset, dimming over the course of the month from mag. 0.3 to 0.5. It ends its retrograde motion on 24 December. ***Uranus*** stays in ***Taurus***, visible shortly after sunset, at mag. 5.6. ***Neptune*** is in Pisces visible from sunset, starting at mag. 7.7 and fading a little to mag. 7.8. It reverts back to direct motion on 15 December.

The path of the Sun and the planets along the ecliptic in December.

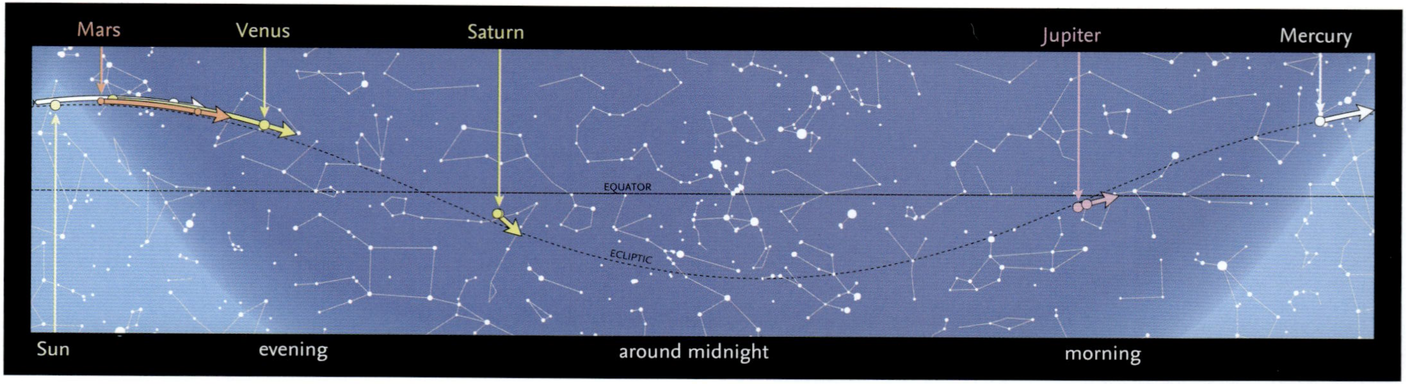

Calendar for December

04	08:42	Moon at apogee = 404,663 km
06	05:22	First Quarter Moon
07		Puppid-Velid meteor shower maximum
12	03:55	Pleiades 0.5°S of the Moon
13	16:09	Full Moon
14		Geminid meteor shower maximum
15	11:13	Neptune ends retrograde motion
16	02:21	Moon at perigee = 364,026 km
18	10:50	Regulus 3.1°N of the Moon
20	09:11	Last Quarter Moon
22	02:43	Winter Solstice
22	08:13	Spica 3.3°N of the Moon
24	03:04	Saturn ends retrograde motion
25	18:01	Antares 0.4°S of the Moon
27	20:12	New Moon
29	12:10	Mars (mag. 1.2) 0.5°S of the Moon
31	01:54	Venus (mag. -4.0) 3.1°S of the Moon

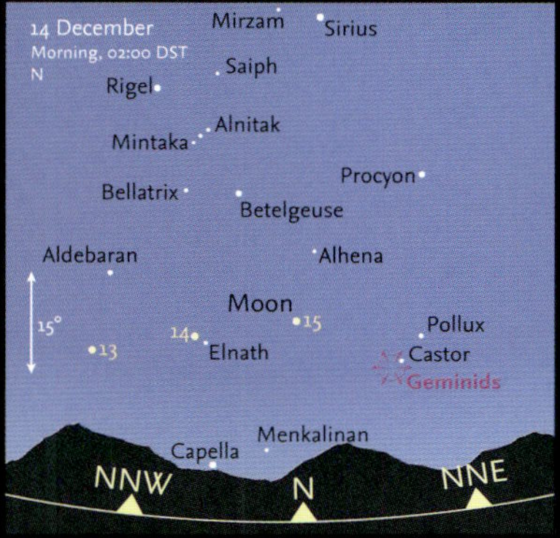

14 December • *The Geminid meteor shower peaks, and moonlight interferes with observations.*

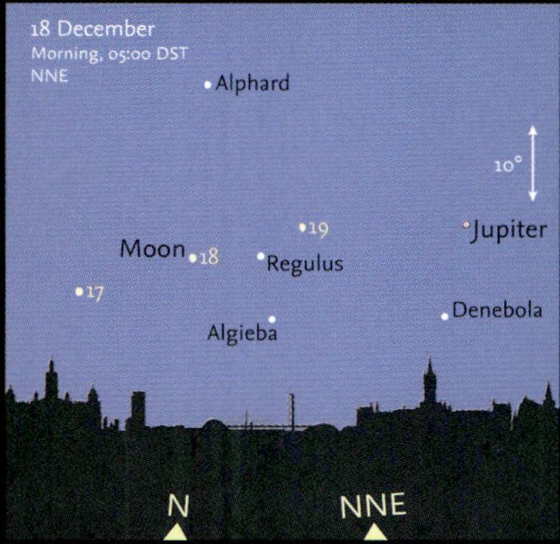

18 December • *Regulus and the waning gibbous Moon.*

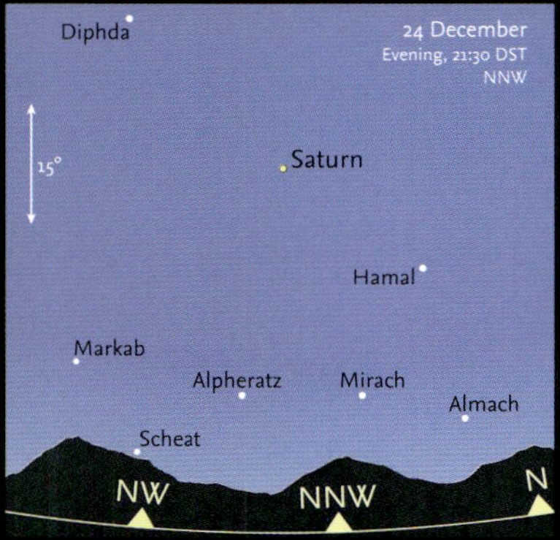

24 December • *Saturn ends retrograde motion.*

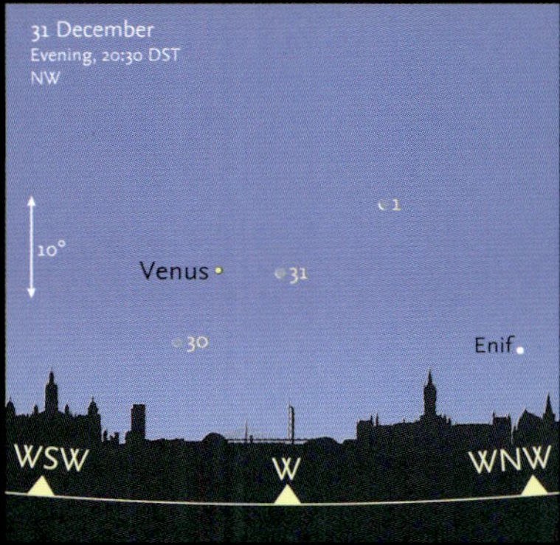

31 December • *Venus and the waxing crescent Moon together. Altair lies to the west.*

Dark Sky Sites

International Dark-Sky Association Sites

The ***International Dark-Sky Association*** (IDA) recognizes various categories of sites that offer areas where the sky is dark at night, free from light pollution and particularly suitable for astronomical observing. Although a number of sites are under consideration, the majority of confirmed sites are in North America. There are at least 13 sites in the southern hemisphere and the list is growing.

Details of the IDA are at: **https://www.darksky.org/**. Information on the various categories and individual sites are at: **https://www.darksky.org/our-work/conservation/idsp/**. New sites are continually being added, so check the IDA website for details. Many of these sites have major observatories or other facilities available for public observing (often at specific dates or times).

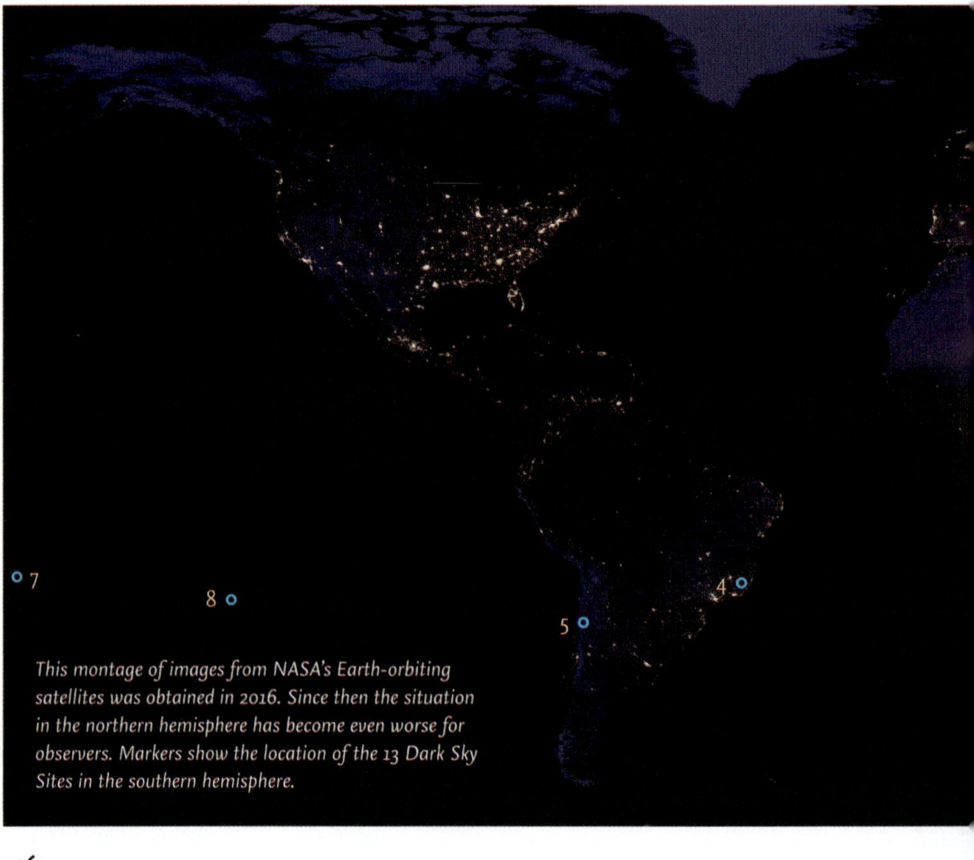

This montage of images from NASA's Earth-orbiting satellites was obtained in 2016. Since then the situation in the northern hemisphere has become even worse for observers. Markers show the location of the 13 Dark Sky Sites in the southern hemisphere.

1 ***!Ae!Hai Kalahari Heritage Park*** (South Africa)
2 ***Aoraki Mackenzie*** (New Zealand)
3 ***Aotea / Great Barrier Island*** (New Zealand)
4 ***Desengano State Park*** (Brazil)
5 ***Gabriela Mistral*** (Chile)
6 ***NamibRand Nature Reserve*** (Namibia)
7 ***Niue*** (New Zealand)
8 ***Pitcairn Islands*** (UK)
9 ***River Murray*** (Australia)
10 ***Stewart Island / Rakiura*** (New Zealand)
11 ***The Jump-Up*** (Australia)
12 ***Wai-Iti*** (New Zealand)
13 ***Warrumbungle National Park*** (Australia)

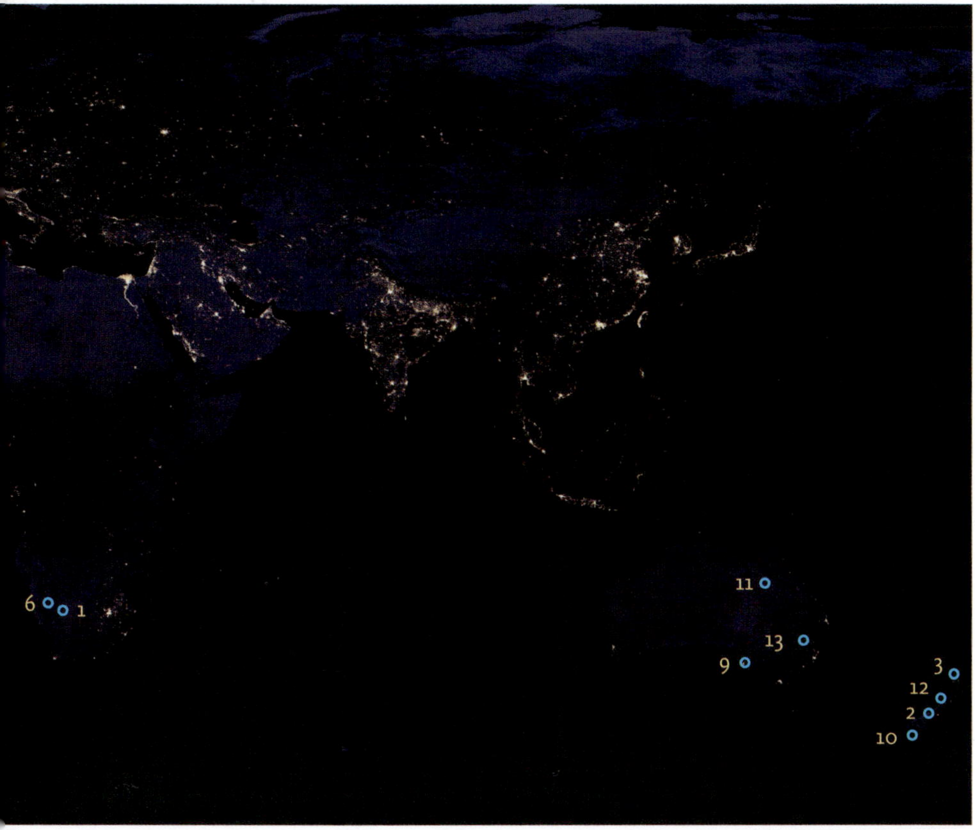

Glossary and Tables

aphelion	The point on an orbit that is farthest from the Sun.
apogee	The point on its orbit at which the Moon is farthest from the Earth.
appulse	The apparently close approach of two celestial objects; two planets, or a planet and star.
astronomical unit	(AU) The mean distance of the Earth from the Sun, 149,597,870 km.
celestial equator	The great circle on the celestial sphere that is in the same plane as the Earth's equator.
celestial sphere	The apparent sphere surrounding the Earth on which all celestial bodies (stars, planets, etc.) seem to be located.
conjunction	The point in time when two celestial objects have the same celestial longitude. In the case of the Sun and a planet, superior conjunction occurs when the planet lies on the far side of the Sun (as seen from Earth). For Mercury and Venus, inferior conjunction occurs when they pass between the Sun and the Earth.
direct motion	Motion from west to east on the sky.
ecliptic	The apparent path of the Sun across the sky throughout the year. Also: the plane of the Earth's orbit in space.
elongation	The point at which an inferior planet has the greatest angular distance from the Sun, as seen from Earth.
equinox	The two points during the year when night and day have equal duration. Also: the points on the sky at which the ecliptic intersects the celestial equator. The vernal (spring) equinox is of particular importance in astronomy.
gibbous	The stage in the sequence of phases at which the illumination of a body lies between half and full. In the case of the Moon, the term is applied to phases between First Quarter and Full, and between Full and Last Quarter.
inferior planet	Either of the planets Mercury or Venus, which have orbits inside that of the Earth.
magnitude	The brightness of a star, planet or other celestial body. It is a logarithmic scale, where larger numbers indicate fainter brightness. A difference of 5 in magnitude indicates a difference of 100 in actual brightness, thus a first-magnitude star is 100 times as bright as one of sixth magnitude.
meridian	The great circle passing through the North and South Poles of a body and the observer's position; or the corresponding great circle on the celestial sphere that passes through the North and South Celestial Poles and also through the observer's zenith.
nadir	The point on the celestial sphere directly beneath the observer's feet, opposite the zenith.
occultation	The disappearance of one celestial body behind another, such as when stars or planets are hidden behind the Moon.
opposition	The point on a superior planet's orbit at which it is directly opposite the Sun in the sky.
perigee	The point on its orbit at which the Moon is closest to the Earth.
perihelion	The point on an orbit that is closest to the Sun.
retrograde motion	Motion from east to west on the sky.
superior planet	A planet that has an orbit outside that of the Earth.
vernal equinox	The point at which the Sun, in its apparent motion along the ecliptic, crosses the celestial equator from south to north. Also known as the First Point of Aries.
zenith	The point directly above the observer's head.
zodiac	A band, streching 8° on either side of the ecliptic, within which the Moon and planets appear to move. It consists of twelve equal areas, originally named after the constellation that once lay within it.

The Constellations

There are 88 constellations covering the whole of the celestial sphere, but 4 of these in the northern hemisphere (Camelopardalis, Cassiopeia, Cepheus and Ursa Minor) can never be seen (even in part) from a latitude of 35°S, so are omitted from this table. The names themselves are expressed in Latin, and the names of stars are frequently given by Greek letters (see next page) followed by the genitive of the constellation name. The genitives and English names of the various constellations are included.

Name	Genitive	Abbr.	English name
Andromeda	Andromedae	And	Andromeda
Antlia	Antliae	Ant	Air Pump
Apus	Apodis	Aps	Bird of Paradise
Aquarius	Aquarii	Aqr	Water Bearer
Aquila	Aquilae	Aql	Eagle
Ara	Arae	Ara	Altar
Aries	Arietis	Ari	Ram
Auriga	Aurigae	Aur	Charioteer
Boötes	Boötis	Boo	Herdsman
Caelum	Caeli	Cae	Burin
Cancer	Cancri	Cnc	Crab
Canes Venatici	Canum Venaticorum	CVn	Hunting Dogs
Canis Major	Canis Majoris	CMa	Big Dog
Canis Minor	Canis Minoris	CMi	Little Dog
Capricornus	Capricorni	Cap	Sea Goat
Carina	Carinae	Car	Keel
Centaurus	Centauri	Cen	Centaur
Cetus	Ceti	Cet	Whale
Chamaeleon	Chamaeleontis	Cha	Chameleon
Circinus	Circini	Cir	Compasses
Columba	Columbae	Col	Dove
Coma Berenices	Comae Berenices	Com	Berenice's Hair
Corona Australis	Coronae Australis	CrA	Southern Crown
Corona Borealis	Coronae Borealis	CrB	Northern Crown
Corvus	Corvi	Crv	Crow
Crater	Crateris	Crt	Cup
Crux	Crucis	Cru	Southern Cross
Cygnus	Cygni	Cyg	Swan
Delphinus	Delphini	Del	Dolphin
Dorado	Doradus	Dor	Dorado
Draco	Draconis	Dra	Dragon
Equuleus	Equulei	Equ	Little Horse
Eridanus	Eridani	Eri	River Eridanus
Fornax	Fornacis	For	Furnace
Gemini	Geminorum	Gem	Twins
Grus	Gruis	Gru	Crane
Hercules	Herculis	Her	Hercules
Horologium	Horologii	Hor	Clock
Hydra	Hydrae	Hya	Water Snake
Hydrus	Hydri	Hyi	Lesser Water Snake
Indus	Indi	Ind	Indian
Lacerta	Lacertae	Lac	Lizard

Name	Genitive	Abbr.	English name
Leo	Leonis	Leo	Lion
Leo Minor	Leonis Minoris	LMi	Little Lion
Lepus	Leporis	Lep	Hare
Libra	Librae	Lib	Scales
Lupus	Lupi	Lup	Wolf
Lynx	Lyncis	Lyn	Lynx
Lyra	Lyrae	Lyr	Lyre
Mensa	Mensae	Men	Table Mountain
Microscopium	Microscopii	Mic	Microscope
Monoceros	Monocerotis	Mon	Unicorn
Musca	Muscae	Mus	Fly
Norma	Normae	Nor	Set Square
Octans	Octantis	Oct	Octant
Ophiuchus	Ophiuchi	Oph	Serpent Bearer
Orion	Orionis	Ori	Orion
Pavo	Pavonis	Pav	Peacock
Pegasus	Pegasi	Peg	Pegasus
Perseus	Persei	Per	Perseus
Phoenix	Phoenicis	Phe	Phoenix
Pictor	Pictoris	Pic	Painter's Easel
Pisces	Piscium	Psc	Fish
Piscis Austrinus	Piscis Austrini	PsA	Southern Fish
Puppis	Puppis	Pup	Stern
Pyxis	Pyxidis	Pyx	Compass
Reticulum	Reticuli	Ret	Net
Sagitta	Sagittae	Sge	Arrow
Sagittarius	Sagittarii	Sgr	Archer
Scorpius	Scorpii	Sco	Scorpion
Sculptor	Sulptoris	Scu	Sculptor
Scutum	Scuti	Sct	Shield
Serpens	Serpentis	Ser	Serpent
Sextans	Sextantis	Sex	Sextant
Taurus	Tauri	Tau	Bull
Telescopium	Telescopii	Tel	Telescope
Triangulum	Trianguli	Tri	Triangle
Triangulum Australe	Trianguli Australis	TrA	Southern Triangle
Tucana	Tucanae	Tuc	Toucan
Ursa Major	Ursae Majoris	UMa	Great Bear
Vela	Velorum	Vel	Sails
Virgo	Virginis	Vir	Virgin
Volans	Volantis	Vol	Flying Fish
Vulpecula	Vulpeculae	Vul	Fox

The Greek Alphabet

α	Alpha	ε	Epsilon	ι	Iota	ν	Nu	ρ	Rho	φ (ϕ)	Phi
β	Beta	ζ	Zeta	κ	Kappa	ξ	Xi	σ (ς)	Sigma	χ	Chi
γ	Gamma	η	Eta	λ	Lambda	ο	Omicron	τ	Tau	ψ	Psi
δ	Delta	θ (ϑ)	Theta	μ	Mu	π	Pi	υ	Upsilon	ω	Omega

Some common asterisms	
Belt of Orion	δ, ε and ζ Orionis
Cat's Eyes	λ and υ Scorpii
Circlet	γ, θ, ι, λ and κ Piscium
False Cross	ε and ι Carinae and δ and κ Velorum
Fish Hook	α, β, δ and π Scorpii
Head of Cetus	α, γ, ξ^2, μ and λ Ceti
Head of Hydra	δ, ε, ζ, η, ρ and σ Hydrae
Job's Coffin	α, β, γ and δ Delphini
Keystone	ε, ζ, η and π Herculis
Kids	ζ and η Aurigae
Milk Dipper	ζ, γ, σ, φ and λ Sagittarii
Pot	= Saucepan
Saucepan	ι, θ, ζ, ε, δ and η Orionis
Sickle	α, η, γ, ζ, μ and ε Leonis
Southern Pointers	α and β Centauri
Square of Pegasus	α, β and γ Pegasi with α Andromedae
Sword of Orion	θ and ι Orionis
Teapot	γ, ε, δ, λ, φ, σ, τ and ζ Sagittarii
Water Jar	γ, η, κ and ζ Aquarii
Y of Aquarius	= Water Jar

Acknowledgements

Our thanks to Gerry Breslin and Samuel Fitzgerald at Collins; and to Ed Bloomer, Bryony Lanigan, Gideon Bendavid-Walker, Jess Lee, Tania de Sales Marques, Patricia Skelton, Carys Jennings, Imo Bell, Greg Brown and Catherine Muller, astronomers at Royal Observatory Greenwich, and Elizabeth Cunningham, Astronomer Emeritus at Royal Observatory Greenwich.

Image Credits

pp. 16, 21, 35, 41, 83, 85 Shutterstock
p. 29 (left) NASA/Johns Hopkins University Applied Physics Laboratory/Carnegie Institution of Washington/Southwest Research Institute
p. 29 (right) Gerald Rhemann
p. 37 ESO/Y Beletsky
pp. 43, 47 Chuck Ayoub
pp. 49, 97 A Fujii (ESA/Hubble)
p. 53 ESO/P Horálek
p. 55 Taavi Niittee/Törva Astronoomiaklubi
p. 59 Nasa/Esa/Hubble
p. 61 ESO/B Tafreshi
pp. 65, 71 ESO/VISTA
p. 67 Vanderbei
p. 73 REU program/NOIRLab/NSF/AURA
p. 77 Arnas Goldberg
p. 79 Ljubinko Jovanovic
p. 89 ESO/VMC Survey
p. 91 ESO
p. 95 Stocktrek Images/Alamy Stock Photo
p. 101 Asim Patel
p. 103 Roberto Mura

Further Information

Books

Bone, Neil (1993), ***Observer's Handbook: Meteors***, George Philip, London & Sky Publ. Corp., Cambridge, Mass.
Chu, A (2012), ***The Cambridge Photographic Moon Atlas***, Cambridge University Press, Cambridge
Dunlop, Storm (1999), ***Wild Guide to the Night Sky***, HarperCollins, London
Dunlop, Storm (2012), ***Practical Astronomy***, 3rd edn, Philip's, London
Dunlop, Storm, Rükl, Antonin & Tirion, Wil (2005), ***Collins Atlas of the Night Sky***, HarperCollins, London
Ellyard, David & Tirion, Wil (2008), ***Southern Sky Guide***, 3rd edn, Cambridge University Press, Cambridge
Heifetz, Milton & Tirion, Wil (2017), ***A Walk through the Heavens***, 4th edn, Cambridge University Press, Cambridge
Heifetz, Milton & Tirion, Wil (2012), ***A Walk through the Southern Sky***, 3rd edn, Cambridge University Press, Cambridge
National Geographic & Wei-Haas, Maya (2023), ***Stargazer's Atlas: The Ultimate Guide to the Night Sky***, National Geographic, Washington, D.C.
O'Meara, Stephen J. (2008), ***Observing the Night Sky with Binoculars***, Cambridge University Press, Cambridge
Ridpath, Ian (2018), ***Star Tales***, 2nd edn, Lutterworth Press, Cambridge
Ridpath, Ian, ed. (2003), ***Oxford Dictionary of Astronomy***, 2nd edn, Oxford University Press, Oxford
Ridpath, Ian, ed. (2004), ***Norton's Star Atlas***, 20th edn, Pi Press, New York
Ridpath, Ian & Tirion, Wil (2004), ***Collins Gem – Stars***, HarperCollins, London
Ridpath, Ian & Tirion, Wil (2017), ***Collins Pocket Guide Stars and Planets***, 5th edn, HarperCollins, London
Ridpath, Ian & Tirion, Wil (2019), ***The Monthly Sky Guide***, 10th edn, Dover Publications, New York
Rükl, Antonín (1990), ***Hamlyn Atlas of the Moon***, Hamlyn, London & Astro Media Inc., Milwaukee
Rükl, Antonín (2004), ***Atlas of the Moon***, Sky Publishing Corp., Cambridge, Mass.
Scagell, Robin (2000), ***Philip's Stargazing with a Telescope***, George Philip, London
Sky & Telescope (2017), ***Astronomy 2018***, Australian Sky & Telescope, Quasar Publishing, Georges Hall, NSW
Stimac, Valerie (2019), ***Dark Skies: A Practical Guide to Astrotourism***, Lonely Planet, Franklin, TN
Tirion, Wil (2011), ***Cambridge Star Atlas***, 4th edn, Cambridge University Press, Cambridge
Topalovic, Radmila & Kerss, Tom (2016), ***Stargazing: Beginner's Guide to Astronomy***, HarperCollins, London

Journals

Astronomy, Astro Media Corp., 21027 Crossroads Circle, P.O. Box 1612, Waukesha, WI 53187-1612 USA. http://www.astronomy.com
Astronomy Now, Pole Star Publications, PO Box 175, Tonbridge, Kent TN10 4QX UK. http://astronomynow.com
Sky at Night Magazine, BBC publications, London. http://skyatnightmagazine.com
Sky & Telescope, Sky Publishing Corp., Cambridge, MA 02138-1200 USA. http://www.skyandtelescope.org/

Societies

British Astronomical Association, Burlington House, Piccadilly, London W1J 0DU. http://www.britastro.org/
The principal British organization for amateur astronomers (with some professional members), particularly for those interested in carrying out observational programmes. Its membership is, however, worldwide. It publishes fully refereed, scientific papers and other material in its well-regarded journal.

Federation of Astronomical Societies, Secretary: Ken Sheldon, Whitehaven, Maytree Road, Lower Moor, Pershore, Worcs. WR10 2NY. http://www.fedastro.org.uk/fas/
An organization that is able to provide contact information for local astronomical societies in the United Kingdom.

Royal Astronomical Society, Burlington House, Piccadilly, London W1J 0BQ. http://www.ras.org.uk/

The premier astronomical society, with membership primarily drawn from professionals and experienced amateurs. It has an exceptional library and is a designated centre for the retention of certain classes of astronomical data. Its publications are the standard medium for dissemination of astronomical research.

Society for Popular Astronomy, 36 Fairway, Keyworth, Nottingham NG12 5DU.

http://www.popastro.com/

A society for astronomical beginners of all ages, which concentrates on increasing members' understanding and enjoyment, but which does have some observational programmes. Its journal is entitled *Popular Astronomy*.

Software

Redshift, Redshift-Live. http://www.redshift-live.com/en/

Starry Night & Starry Night Pro, Sienna Software Inc., Toronto, Canada. http://www.starrynight.com

Stellarium, https://stellarium.org/

Internet sources

There are numerous sites with information about all aspects of astronomy, and all of those have numerous links. Although many amateur sites are excellent, treat any statements and data with caution. The sites listed below offer accurate information. Please note that the URLs may change. If so, use a good search engine, such as Google, to locate the information source.

Information

Astronomical data (inc. eclipses) HM Nautical Almanac Office: http://astro.ukho.gov.uk

Auroral information Michigan Tech: http://www.geo.mtu.edu/weather/aurora/

American Meteor Society: http://amsmeteors.org/

Comets JPL Solar System Dynamics: http://ssd.jpl.nasa.gov/

Deep-sky objects Saguaro Astronomy Club Database: http://www.virtualcolony.com/sac/

Eclipses NASA Eclipse Page: http://eclipse.gsfc.nasa.gov/eclipse.html

Ice in Space (Southern Hemisphere Online Astronomy Forum): http://www.iceinspace.com.au/index.php?home

Moon (inc. Atlas) Inconstant Moon: http://www.inconstantmoon.com/

Planets Planetary Fact Sheets: http://nssdc.gsfc.nasa.gov/planetary/planetfact.html

Satellites (inc. International Space Station)

Heavens Above: http://www.heavens-above.com/

Visual Satellite Observer: http://www.satobs.org/

Star Chart http://www.skyandtelescope.com/observing/interactive-sky-watching-tools/interactive-sky-chart/

What's Visible

Skyhound: http://www.skyhound.com/sh/skyhound.html

Skyview Cafe: http://www.skyviewcafe.com

Institutes and Organizations

European Space Agency: http://www.esa.int/

International Dark-Sky Association: http://www.darksky.org/

Jet Propulsion Laboratory: http://www.jpl.nasa.gov/

Lunar and Planetary Institute: http://www.lpi.usra.edu/

National Aeronautics and Space Administration: http://www.hq.nasa.gov/

Solar Data Analysis Center: http://umbra.gsfc.nasa.gov/

Space Telescope Science Institute: http://www.stsci.edu/